Microwear Patterns on Experimental Basalt Tools

Thomas H. Richards

BAR International Series 460
1988

B.A.R.

5, Centremead, Osney Mead, Oxford OX2 0DQ, England.

BAR -S460, 1988: 'Microwear Patterns on Experimental Basalt Tools'

ISBN 9780860545910 paperback
ISBN 9781407347462 e-book
DOI https://doi.org/10.30861/9780860545910
A catalogue record for this book is available from the British Library
This book is available at www.barpublishing.com

To Sharon Mueller

and

my parents

Waltraud and Tom Richards

CONTENTS

LIST OF FIGURES v

LIST OF TABLES xi

PREFACE xiii

Chapter

1. INTRODUCTION

1.1 Research Goals 1
1.2 Thesis Outline 2

2. PREVIOUS MICROWEAR RESEARCH

2.1 The Current State of Experimental Microwear Research 4
2.2 Previous Microwear Research Involving Basalt Tools 7

3. GEOLOGY AND PETROGRAPHY OF CACHE CREEK BASALT

3.1 Introduction 10
3.2 Geomorphology of the Cache Creek Basalt Source Area 10
3.3 Geological Origin and Petrography 14
3.4 Petrographic Analysis 15

4. HISTORIC AND PREHISTORIC USE OF CACHE CREEK BASALT

4.1 Introduction 24
4.2 Thompson, Shuswap and Lillooet Lithic Technology 26
4.3 Prehistoric Use of Basalt 29
4.4 Summary and Discussion 37

5. INDEPENDENT VARIABLES, RECORDING METHODS, AND PROCEDURES

5.1 Introduction 38
5.2 Pre-Experiment Observations 48
5.3 Observations During Experiments 51

6. DEPENDENT VARIABLES, RECORDING METHODS, AND PROCEDURES

6.1 Introduction 59
6.2 Analytical Procedure 59
6.3 Recording Methods 60
6.4 Equipment 60
6.5 Dependent Variables 61
6.7 Natural "Wear 66

7. MICROWEAR OBSERVED ON EXPERIMENTAL BASALT TOOLS

7.1 Introduction 69
7.2 Plants 70
7.3 Hide 71
7.4 Bone 76
7.5 Antler 80
7.6 Fish Butchering 84
7.7 Deer Butchering 87
7.8 Wood 93

8. STATISTICAL ANALYSIS OF MICROFLAKE SCAR VARIABLES

8.1 Introduction 101
8.2 Correlations Between Independent and Dependent Variables 101
8.3 Cluster Analysis 108
8.4 Discriminant Function Analysis 114

9. BLIND TESTS

9.1 Introduction 121
9.2 Background 122
9.3 Discussion of Blind Test Results 122
9.4 Summary and Discussion 132

10. RELATIONSHIPS BETWEEN TOOL-USE BEHAVIOUR AND MICROWEAR ON EXPERIMENTAL BASALT TOOLS

10.1 Introduction 133
10.2 Vitreous Basalt Microwear Summary 133
10.3 Non-Vitreous Basalt Microwear Summary 146
10.4 Scanning Electron Microscope Study Summary 147

11. SUMMARY AND CONCLUSIONS

11.1 Summary of Research Conclusions 149
11.2 Future Research Directions 152

REFERENCES 154

APPENDIX A: Tabulations of Microwear Data 168

APPENDIX B: Photomicrographs of Microwear 202

APPENDIX C: SEM Photomicrographs of Microwear 222

APPENDIX D: Drawings of Experimental Tools 251

APPENDIX E: Blind Test Interpretations 272

LIST OF FIGURES

Figure Page

1. Southern British Columbia 11

2. Location of basalt source 12

3. Basalt cobble in alluvial deposit 13

4. Basalt cobble in till 13

5. Thin section of cobble 22 19

6. Thin section of cobble 44 19

7. Thin section of cobble 21 20

8. Thin section of cobble 21 20

9. Thin section of cobble 25 21

10. Thin section of cobble 35 21

11. Thin section of cobble 26 22

12. Thin section of cobble 26 22

13. Thin section of cobble 34 23

14. Thin section of cobble 34 23

15. Ethnographic map of British Columbia 25

16. Slight pre-use damage 49

17. Severe pre-use damage 49

18. Tool 134-23 being used to scrape wood 52

19. Close-up of tool 134-23 in use 52

20. Tool 27-16 being used to scrape hide 53

21. Close-up of tool 46-4 during use as scraper 53

22. Tool 12-11 being used to butcher deer 54

23. Tool 1-4 being used to butcher deer 54

24. Tool actions 56

25. Dorsal and ventral edge-aspects 62

26. Microflake scar relationships 63

27. Leading and trailing edge-aspects 65

28. Orientation of striations 67

29. Cluster analysis dendrogram 111

30. Model of polish stage development 138

31. Rates of polish development 141

32. Unused tool 134-23 203

33. Unused tool 27-17 203

34. Microwear on tool 134-11 204

35. Microwear on tool 134-11 204

36. Microwear on tool 161-11 205

37. Microwear on tool 17-14 205

38. Microwear on tool 17-14 206

39. Microwear on tool 12-13 206

40. Microwear on tool 12-13 207

41. Microwear on tool 27-17 207

42. Microwear on tool 27-17 208

43. Microwear on tool 27-17 208

44. Microwear on tool 134-9 209

45. Microwear on tool 134-9 209

46. Microwear on tool 46-2 210

47. Microwear on tool 46-2 210

48. Microwear on tool 46-2 211

49. Microwear on tool 46-2 211

50. Microwear on tool 17-13 212

51. Microwear on tool 17-13 212

52. Microwear on tool 17-13 213

53. Microwear on tool 138-19 213

54. Microwear on tool 134-6 214

55. Microwear on tool 134-6 214

56. Microwear on tool 134-6 215

57. Microwear on tool 141-20 215

58. Microwear on tool 141-20 216

59. Microwear on tool 141-20 216

60. Microwear on tool 141-20 217

61. Microwear on tool 141-20 217

62. Microwear on tool 12-3 218

63. Microwear on tool 45-7 218

64. Microwear on tool 27-5 219

65. Microwear on tool 27-5 219

66. Microwear on tool 134-23 220

67. Microwear on tool 134-23 220

68. Microwear on tool 134-23 221

69. Microwear on tool 134-23 221

70. Unused basalt tool (SEM) 223

71. Unused basalt tool (SEM) 223

72. Microwear (SEM) on tool 12-12 224

73. Microwear (SEM) on tool 12-12 224

74. Microwear (SEM) on tool 12-12 225
75. Microwear (SEM) on tool 17-14 225
76. Microwear (SEM) on tool 17-14 226
77. Microwear (SEM) on tool 17-14 226
78. Microwear (SEM) on tool 17-14 227
79. Microwear (SEM) on tool 17-14 227
80. Microwear (SEM) on tool 27-17 228
81. Microwear (SEM) on tool 27-17 228
82. Microwear (SEM) on tool 27-17 229
83. Microwear (SEM) on tool 27-17 229
84. Microwear (SEM) on tool 134-9 230
85. Microwear (SEM) on tool 134-9 230
86. Microwear (SEM) on tool 134-9 231
87. Microwear (SEM) on tool 46-2 231
88. Microwear (SEM) on tool 18-19 232
89. Microwear (SEM) on tool 18-19 232
90. Microwear (SEM) on tool 17-13 233
91. Microwear (SEM) on tool 17-13 233
92. Microwear (SEM) on tool 134-22 234
93. Microwear (SEM) on tool 134-22 234
94. Microwear (SEM) on tool 17-11 235
95. Microwear (SEM) on tool 18-13 235
96. Microwear (SEM) on tool 161-19 236
97. Microwear (SEM) on tool 138-12 236
98. Microwear (SEM) on tool 134-17 237

99. Microwear (SEM) on tool 134-17 237
100. Microwear (SEM) on tool 12-6 238
101. Microwear (SEM) on tool 12-6 238
102. Microwear (SEM) on tool 12-6 239
103. Microwear (SEM) on tool 12-6 239
104. Microwear (SEM) on tool 45-7 240
105. Microwear (SEM) on tool 45-7 240
106. Microwear (SEM) on tool 12-11 241
107. Microwear (SEM) on tool 12-11 241
108. Microwear (SEM) on tool 44-8 242
109. Microwear (SEM) on tool 44-8 242
110. Microwear (SEM) on tool 12-1 243
111. Microwear (SEM) on tool 12-1 243
112. Microwear (SEM) on tool 141-6 244
113. Microwear (SEM) on tool 141-6 244
114. Microwear (SEM) on tool 141-6 245
115. Microwear (SEM) on tool 141-17 245
116. Microwear (SEM) on tool 141-17 246
117. Microwear (SEM) on tool 27-15 246
118. Microwear (SEM) on tool 27-15 247
119. Microwear (SEM) on tool 27-15 247
120. Microwear (SEM) on tool 27-15 248
121. Microwear (SEM) on tool 27-15 248
122. Microwear (SEM) on tool 134-18 249
123. Microwear (SEM) on tool 134-18 249

124. Microwear (SEM) on tool 12-5 250
125. Microwear (SEM) on tool 12-5 250
126. Plant cutting tools 252
127. Hide working tools 253
128. Hide working tools 254
129. Hide working tools 255
130. Bone working tools 256
131. Bone working tools 257
132. Antler working tools 258
133. Antler working tools 259
134. Antler working tools 260
135. Antler working tools 261
136. Fish butchering tools 262
137. Deer butchering tools 263
138. Deer butchering tools 264
139. Deer butchering tools 265
140. Deer butchering tools 266
141. Wood working tools 267
142. Wood working tools 268
143. Wood working tools 269
144. Wood working tools 270
145. Wood working tools 271

LIST OF TABLES

Table Page

1. Summary of basalt microwear research 8
2. Basalt thin section descriptions 17
3. List of mammal species 30
4. List of fish species 31
5. List of plant and tree species 32
6. List of organic material culture items 34
7. List of chipped stone tool types 36
8. Experimental tool pre-use attributes 39
9. Summary of tool use experiments 44
10. Dependent variable definitions 102
11. Independent variable definitions 104
12. Spearman Correlations 106
13. Spearman Correlations using percentages 107
14. Agglomeration methods and distance measures 109
15. Dependent variable distribution summary 113
16. Dependent variable sets 113
17. Correctly classified tools 117
18. Correlations with discriminant functions 118
19. Blind Test tool classification results 119
20. Summary of Blind Test tool use 123
21. Three-part contact material scale 126
22. Four-part contact material scale 126
23. Blind Test 1 results 127

24. Blind Test 2 results 129

25. Blind Test 3 results 131

26. Polish development stage descriptions 139

27. Summary of microwear patterns 143

28. Microflake scar distributions 169

29. Micropolish characteristics 177

30. Striation characteristics 190

PREFACE

This book is based on a thesis submitted in partial fulfillment of the requirements for a Master of Arts degree in Anthropology and Archaeology at the University of Saskatchewan (Richards 1987). The Department of Anthropology and Archaeology generously supported my research with graduate scholarships, sessional lectureships, and teaching assistantships.

The members of my committee provided insightful and helpful comments on my thesis: Dr. James F.V. Millar (supervisor and chairman), Dr. Ahmet R. Mermut, Dr. Ernest G. Walker, Mr. Satya P. Sharma, and Mr. John G. McConnell (external examiner). I owe a special debt to Dr. Mermut for his encouragement, for guiding me through the thin section and scanning electron microscope analyses, and for providing supplies, equipment and his time so freely. Professor Millar, my supervisor, did everything in his power to ensure the successful completion of the research project. He was always available for consultation and advice, and he ensured that I never strayed too far from my topic.

I thank the following people for their contributions to my research: Leslie Amundson, John Brandon, Joanne Hader, Sharon Hanna, Eldon Johnson, David Kelly, Kit Krozser, Kathy Lipsett and Ginger Scanlon, my fellow grad students at the University of Saskatchewan who helped in a variety of ways, Geordie Howe (Simon Fraser University) for participating in the blind tests, Don Glossop and Don Florizone (Saskatchewan Research Council) for providing access to an incident light microscope, and Glen Hudy (University of Saskatchewan) for his invaluable assistance during the preparation of thin sections and during the scanning electron microscope analysis.

The Department of Archaeology and Prehistory, University of Sheffield, have generously provided facilities for the production of the manuscript of this book. In particular, I thank Dorothy Cruse, Mavis Torry, Dr. Patricia Phillips and Professor Keith Brannigan.

Thomas Richards

Sheffield, November 1988

CHAPTER 1

INTRODUCTION

1.1 *Research Goals*

Archaeologists are increasingly concerned with reconstructing prehistoric cultural behaviour to attain a better understanding of past societies. Stone tools have a high potential for contributing to behavioural reconstruction due to the wealth of technological, functional and ideological information they embody, and archaeologists face the formidable task of reliably retrieving this information. Binford (1981,1982) advocates the development of "middle range theory" to provide a means whereby material remains can be related to the past behaviour which produced them. The present study is directed at understanding the relationships between tool-use behaviour (or function) and microwear patterns on Cache Creek basalt tools. The problem is explored through controlled tool-use experimentation and the observation of microwear traces on used tools by means of light microscopes and a scanning electron microscope.

The primary goal of this study is to develop a framework for the functional interpretation of basalt tools from late prehistoric archaeological sites located in the southwestern interior of British Columbia. Historically, this area was occupied by the Interior Salish speaking Thompson, Lillooet (Fraser Canyon Lillooet) and Shuswap people of the Plateau culture area. In the 19th century these societies consisted of egalitarian hunter-gatherers who lived in semi-permanent pithouse villages along the major salmon-bearing rivers and lakes. Archaeological research has led to the recognition of prehistoric cultures with similar lifeways occupying the area for at least the past 3500-4000 years. "Cache Creek basalt" is known to have been the primary raw material used for making chipped stone tools by both the historic and prehistoric Native people occupying this area. The ethnographic record and archaeological evidence are used to provide a systemic framework for the tool-use experiments in this study.

Previous microwear research with basalt tools (see Chapter 2.2) has left many unanswered questions which the present study will investigate. The following specific problems are addressed in the course of exploring the relationships between microwear and tool function:

1. Do vitreous and non-vitreous basalts have similar microwear characteristics?

2. Which types of microwear are most useful for identifying the used area of a tool, tool action and contact material?

3. Which magnifications are most appropriate for functional interpretations of basalt tools?

4. Do distinctive types of polish develop on basalt tools as a result of use on different contact materials as has been found to be the case with various flints and cherts?

5. Are rarely observed wear types such as polish and striations more commonly observed when high magnifications are employed?

6. Are multivariate statistical techniques appropriate for the functional classification of basalt tools (and stone tools in general), and can statistical methods be used to identify functionally significant wear variables.?

1.2 *Outline*

In Chapters 2, 3 and 4 pertinent background information to the research is presented. Chapter 2 is a review of experimental microwear research, and contains a section dealing specifically with previous investigations into the microwear characteristics of various basalts. The characteristics of Cache Creek basalt are discussed in Chapter 3, including geology, petrography and distribution. Ethnographic and archaeological data are presented in Chapter 4 to provide a systemic context for the tool-use experiments.

Chapters 5, 6, and 7 pertain to the tool-use experiments. The independent variables are defined and experimental procedures are described in Chapter 5. The dependent variables are defined in Chapter 6. Chapter 7 is the descriptive heart of this study, and consists primarily of descriptions of microwear traces observed on the experimental tools. Four appendices to Chapter 7 contain photomicrographs of the observed microwear, detailed tabulations of microwear on each tool, and illustrations of the experimental tools.

Chapters 8, 9, and 10 are concerned with analysis of the microwear observed on the experimental tools and it is within these chapters that the research problems are addressed and the interpretive methods presented. Chapter 8 consists of statistical analyses of selected microwear variables designed to assess the appropriateness of two multivariate methods and the usefulness of microflake scar attributes for functional analysis. Chapter 9 presents the results of a series of "blind tests" used to point out the strengths and weaknesses in the preliminary interpretive framework. Valuable "lessons" learned in the course of these tests are used to refine and improve the methods presented in final form in Chapter 10. In Chapter 10 the microwear observations, blind test results and statistical analyses are integrated to elucidate the relationships between microwear and tool-use behaviour.

Chapter 11 is a summary of the research results and conclusions, with a discussion of how the research goals were addressed. Additionally, directions for future research are suggested.

CHAPTER 2

PREVIOUS MICROWEAR RESEARCH

2.1 *The Current State of Experimental Microwear Research*

The study of microscopic wear traces on stone tools consists of both experimental and applied research. The use of methods developed through experimentation to determine the function of stone tools from archaeological sites is the applied aspect of microwear research. Experimental microwear research mainly involves making and using stone tools, studying the resulting microwear traces, and developing methods for interpreting microwear on prehistoric tools. "Ethnographic" microwear research is closely allied to experimental research, and involves the study of microwear traces on stone tools used by aboriginal peoples in recorded contexts (see Gould, Koster and Sontz 1971; Hayden 1979b; Nissen and Dittemore 1974; Hester and Follett 1976). There is also a theoretical aspect to experimental microwear research which is directed at achieving a general understanding of how different wear types are formed, and involves striation and polish formation theories and fracture mechanics (see Odell 1981; Del Bene 1979; Lawn and Marshall 1979; Diamond 1979; Cotterell and Kamminga 1979; Tsirk 1979; Kamminga 1978,1979; Unger-Hamilton 1983,1984; Meeks et al. 1982).

Experimental microwear researchers employ a variety of magnifications, with the effective limit of light microscopes being about 600 x. Some researchers use the full range of magnifications available, while others employ either only high magnifications (generally 200 x to 400 x) or low magnifications (below 100 x). The magnifications employed have far reaching consequences regarding the types of wear observed, and therefore, on the entire methodology developed from the research. There has been considerable discussion on the relative merits of using either high or low magnification methodologies (see Holley and

Del Bene 1981; Keeley 1974,1981; Odell 1975,1982; Odell and Odell-Vereecken 1980; Newcomer, Grace and Unger-Hamilton 1986; Richards 1984; Bamforth 1986). Some researchers have also taken advantage of the very high magnifications, enormous depth of field and excellent photomicrographs provided by the scanning electron microscope (see Anderson 1980; Del Bene and Shelley 1979; Del Bene 1979; Fedje 1979; Hay 1977; Lawrence 1979; Shiner and Porter 1974; Unger-Hamilton 1984).

Experimental microwear studies can be subdivided into those which rigidly control the variables, known as "analytic" (Stafford and Stafford 1983) or "mechanical" (Schutt 1982) experiments and those which strive to recreate aboriginal conditions, sacrificing some degree of control, known as "systemic" (Stafford and Stafford 1983) or "goal oriented" (Schutt 1982) experiments. Applicability of the results of experiments to the prehistoric situation they are intended to illuminate is dependent on the degree to which the experimental conditions can duplicate the prehistoric situation. However, some experiments must be rigidly controlled and undertaken somewhat artificially, especially if the researcher is attempting to determine the relationships between specific independent and dependent variables. Most experimental programs designed to produce wear patterns to be used for the interpretation of prehistoric tools sacrifice some authenticity in order to maintain a degree of control over the variables. One study designed to test the differences between wear patterns produced by "mechanical" and "goal oriented" experiments (Schutt 1982) showed only a few moderately significant differences in the wear patterns. As one might expect, the "mechanical" experiments produced wear patterns which were clear and relatively unambiguous, while the "goal oriented" experiments produced wear patterns which were slightly less distinctive. However, Schutt's study is severely flawed because the key variable of lithic raw material was not kept constant between the two series of experiments.

Some microwear studies deal with one tool type, such as scrapers or burins, while others comprise well-rounded experimental frameworks including many tool actions and contact materials (e.g. Keeley 1980; Tringham et al 1974; Odell 1977; Kamminga 1978; Vaughan 1985). It is the extensive experimental frameworks which have formed the basis for advances in knowledge in this field in recent years and allowed innovative applied research to be undertaken (e.g. Keeley 1977,1980; Keeley and Toth 1981; Cahen, Keeley and Van Noten 1979; Odell 1977,1980a; Yerkes 1983).

Researchers have identified a number of types of microwear of which the most useful for determining tool function are polishes, microflake scars, striations, and edge rounding. Certain researchers have become identified with one type of wear: Semenov (1964) is noted for bringing attention to microscopic striations as an indicator of tool motion; Tringham and Odell (Tringham et al. 1974; Odell 1977; Odell and Odell-Vereecken 1980) are well known for providing a broad experimental basis for the use of microflake scars for determining tool action and also relative contact material hardness; and Keeley (1977,1980) has been in the forefront of championing the usefulness of micro-polishes for determining contact materials.

Experimentation with stone tools has involved a wide variety of actions, in fact virtually every type of action described in the ethnographies and those suspected to have been used prehistorically. These actions include sawing, slicing, cutting, scraping, planing, whittling, paring, drilling, boring, perforating, engraving, chopping, adzing, abrading, pounding and use as a projectile tip. The most heavily studied action is scraping, which was dealt with in all of the major experimental programmes as well as in several small scale studies (e.g. Brink 1978a; Broadbent and Knutsson 1975; Price-Beggerly 1976; Siegel 1984; Levitt 1976; Hayden 1979b; Nissen and Dittemore 1974).

A great variety of materials ("contact materials") have been worked with experimental stone tools, including bone, cartilage, fish scales, turtle shell, sea shells, antler, hide, leather, hair, flesh, fat, wood, bark, wild and domesticated plants, vegetables, plant fibers, earth and stone. Contact materials have been worked in their fresh, natural state or they have been modified in ways likely to have occurred in prehistory. The materials have had grit, fat or grease added to them, and have also been soaked in water, boiled in water, cooked, frozen, heated, dried, and cleaned. Again, the large scale experimental programmes have used a variety of contact materials appropriate to the prehistoric situation for which a study was targeted, while smaller scale studies have employed only one or a few contact materials.

A wide range of lithic raw materials have been used in experimental research, including *flint* (Keeley 1980; Tringham et al. 1974; Odell 1977; Vaughan 1981, 1985; Anderson 1980a; Buller 1983; Moss 1983a,b; Unger-Hamilton 1983,1984b,1985; White 1982), *chert* (Brink 1978a,b; Brose 1975; Lawrence 1979; Lewenstein 1984; Kamminga 1978; Lynott 1975; McConaughy 1979; Sabo 1982; Yerkes 1983), *chalcedony* (Lewenstein 1984; Greiser and Sheets 1979; Kamminga 1978), *obsidian* (Hay 1977; Fedje 1979; Kamminga 1978; Corruccini 1985; Flenniken and Haggarty 1979; Greiser and Sheets 1979; Hester et al 1976; Lewenstein 1981,1984; Vaughan 1981), *basalt* (Price-Beggerly 1976; Kamminga 1978; Montgomery 1978; Wyant and Bayham 1976; Odell 1980b; Stafford 1977; Schutt 1982), *quartzite* (Toll 1978; Kamminga 1978; Greiser and Sheets 1979), *quartz* (Broadbent and Knutsson 1975; Sussman 1984,1985; Kamminga 1978), *volcanic tuff* (Kamminga 1978) and *rhyolite* (Roberts 1980; Foix and Bradley 1985). It is apparent that most experimental microwear research has been undertaken with either European flint (and other flints, cherts and chalcedonies with similar physical and microwear properties) or obsidian. When discussing the achievements of microwear research, one is usually referring to studies employing these materials, especially the flints. The applicability of methods developed for flint to other lithic raw materials must be determined through experimentation on a case by case basis.

The accuracy of microwear interpretations has been demonstrated by blind tests of the two major methodologies developed in the 1970s. The high magnification approach developed by Keeley was tested first and proved very effective in identifying the used part of a tool and determining how a tool was used, but was somewhat less accurate in identifying which contact material the tool was used on (Keeley and Newcomer 1977; Newcomer and Keeley 1979). A second test of this methodology achieved even better results (Gendel and Pirnay 1982), however, a recent series of blind tests has thrown the validity of this method into doubt (Newcomer, Grace and Unger-Hamilton 1986). The low magnification approach developed by Tringham and Odell also tested out very well, being equally effective in identifying the used part of a tool, how it was used, and the relative hardness of a contact material, although this methodology proved ineffective for identifying specific contact materials (Odell and Odell-Vereecken 1980).

On the frontiers of microwear research are attempts to eliminate subjectivity in identifying tool function by measuring or quantifying wear (e.g. Semenov and Shchelinskii 1971; Keeley 1977,1980; Grace, Graham and Newcomer 1985; Tomenchuk 1979,1983,1984; Dumont 1982). This research has not yet resulted in practical or widely applied techniques. Other emerging directions involve microscopic and chemical analyses of residues on stone tools to determine function. Researchers have detected microscopic plant phytoliths (Anderson 1980), dried blood (Loy 1983), amino acids (Broderick 1979) and other residues (Shafer and Holloway 1979; Briuer 1976; Paull 1984; Broderick 1982) indicative of contact materials.

2.2 *Previous Microwear Research Involving Basalt Tools*

As is evident in section 2.1, the majority of recent experimental microwear studies are concerned with the development of wear patterns on flints, cherts and obsidian while lesser efforts have been directed towards "coarser" lithic materials, including quartzite, quartz, basalt and rhyolite. Of this group basalt has received the most attention, but no major experimental investigation has been made into its specific use-wear characteristics. Previous experimental microwear studies involving basalt are summarized on Table 1 and discussed below.

With the amount of research undertaken to date employing basalt tools (see Table 1) it would seem reasonable to expect that significant conclusions or generalizations could be made about the wear patterns which develop on basalt. Unfortunately, very few generalizations can be made, for four main reasons: (1) varying qualities of basalt (i.e. coarse-grained to glassy) have been employed by different researchers, hindering close comparisons of wear patterns because the texture of basalt affects the types of wear developed and its visibility under a microscope; (2) researchers used different magnifications while observing wear, and in microwear research the magnifications employed largely determine which types of wear can be observed; (3) experimentally-derived wear patterns were not all equally well illustrated, nor were they described using a consistent terminology; and, (4) the experiments often involved different actions, contact materials, and amounts of tool-use. As an example of the non-comparability of these experiments, Wyant and Bayham (1976) used medium-grained basalt tools for butchering rabbits and sawing or scraping wood and bone. The tools were used for only 100 strokes each, and studied under a microscope at 30 x. In

Table 1. Summary of previous experimental microwear research involving basalt tools.

Reference	Summary
Kamminga 1978	basalt: altered, andesitic and olivine basalts from Australia; actions: scraping, engraving, cutting, perforating, sawing, drilling; contact materials: hide, bone, flesh, turtle shell, sea shell, wood, potatoe; duration of use: until final or stable wear patterns were developed; number of tools and retouch: 44 unretouched; magnification employed: 50 x; types of wear observed: microflake scars, edge blunting, edge dulling, edge rounding, polish, smoothing, abrasion, striations.
Montgomery 1978	basalt: vitreous from Cache Creek, B.C.; actions: paring, scraping, sawing; contact materials: antler, bone, wood, flesh; duration of use: 50 to 400 strokes; number of tools and retouch: 17 unretouched; magnifications employed: 14 x; types of wear observed: microflake scars, edge rounding, crushing.
Odell 1980b	basalt: fine-grained from Cache Creek, B.C.; actions: slicing, sawing; contact materials: flesh, hide, bone, sinew, fat; duration of use: 450 to 950 strokes; number of tools and retouch: 9 unretouched; magnifications employed: 10-80 x; type of wear observed: microflake scars.
Price-Beggarly 1976	basalt: fine-grained from Oahu, Hawaii; actions: scraping; contact materials: wood; duration of use: 250 to 500 strokes; number of tools and retouch: 16 unifacially retouched; magnification employed: 8 x; type of wear observed: edge blunting.
Schutt 1982	basalt: vitreous from White Rock Canyon, New Mexico; actions: scraping, sawing, whittling; contact materials: wood, bone, leather; duration of use: 200 to 2310 strokes; number of tools and retouch: 31 unretouched; magnifications employed: 30-70 x; types of wear observed: microflake scars, edge rounding, polish.
Stafford 1977	basalt: fine-grained from Ashfork, Arizona; actions: scraping, graving; contact materials: wood, bone; duration of use: 100 strokes; number of tools and retouch: 120 burinated; magnifications employed: low; types of wear observed: microflake scars, edge crushing.
Wyant and Bayham 1976	basalt: medium-grained metabasalt from Arizona; actions: cutting, scraping; contact materials: wood, bone, flesh; duration of use: 100 strokes; number of tools and retouch: 17 unretouched; magnification employed: 30 x; types of wear observed: microflake scars, edge regularization.

contrast, Odell (1980b) employed fine-grained basalt tools to butcher a raccoon and a porpoise, with individual tools used between 300 and 1500 strokes. He studied the edges using magnifications between 10 x and 80 x.

The following observations regarding microwear on basalt tools can be made, keeping in mind the above problems and one more limitation. Since all of the research used low magnifications, some types of wear (i.e. polish and striations) may not have been observed simply because the magnifications used were too low to see them. Microflake scars are the primary wear type recorded in all but one of the experimental studies employing basalt. Most researchers concentrate on the size, shape, termination, distribution and patterning of microflake scars, almost to the exclusion of other types of wear (e.g. Odell 1980b; Stafford 1977; Wyant and Bayham 1976; Montgomery 1978). Striations have rarely been observed on basalt tools (e.g. Odell and Odell-Vereecken 1980; Schutt 1982; Kamminga 1978) and polish has only been commonly observed by one researcher (Schutt 1982). Edge-rounding was quite variable in its occurrence, being very common in some experimental assemblages (e.g. Price-Beggerly 1976; Shutt 1982).

It is clear that much remains to be learned about the microwear properties of basalts. The present research is directed at filling in some of the gaps in our knowledge, especially regarding Cache Creek basalt. The geological origin and petrographic characteristics of this basalt are described in the following chapter.

CHAPTER 3

GEOLOGY AND PETROGRAPHY OF CACHE CREEK BASALT

3.1 *Introduction*

The basalt used in this study came from lower Cache Creek, just east of the village of Cache Creek, British Columbia (Figures 1,2). It was obtained from an area approximately 2.5 km upstream of the confluence of Cache Creek with the Bonaparte River, and about 1 km downstream of the point where the creek emerges from between the Arrowstone Hills and the Cache Creek Hills. Basalt was collected from about a 1 km stretch of the creek and within 100 m of the creek. It was in the form of sub-angular to rounded pebbles, cobbles and boulders .5 m or more in length. This source-area for basalt is well known to archaeologists in British Columbia (Richards and Rousseau 1987:34).

3.2 *Geomorphology of the Cache Creek Basalt Source Area*

Bounding lower Cache Creek and its narrow alluvial plain are till deposits from the Fraser glaciation (Ryder 1976). The till, which overlies bedrock, is comprised of a heterogenous mixture of particles ranging from clay to boulders derived from the local bedrock (Ryder 1976:3). In places the till is very thin (less than 2 m thick) and bedrock outcrops just north of the creek. The whole area, discounting recent disturbances, is capped with a thin veneer of aeolian sand and silt (Ryder 1976:3). Basalt can be found in the creek and floodplain deposits, as well as in the adjacent till (Figures 3,4). Recent human disturbances have caused the removal of soil and aeolian deposits near the creek in the collection area, providing large exposures of till and easy access to basalt.

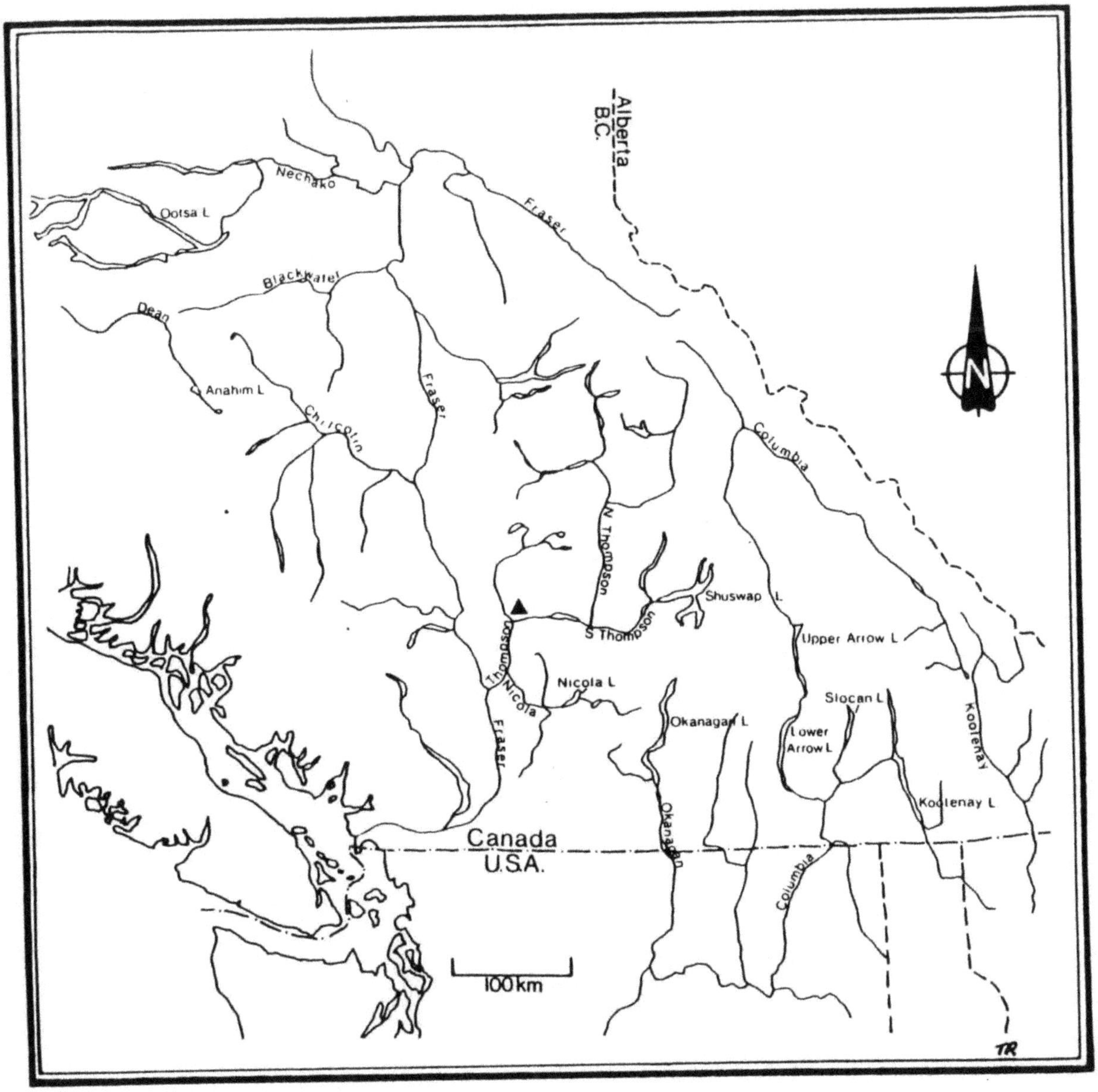

Figure 1. Southern British Columbia with the Cache Creek basalt source marked by a black triangle.

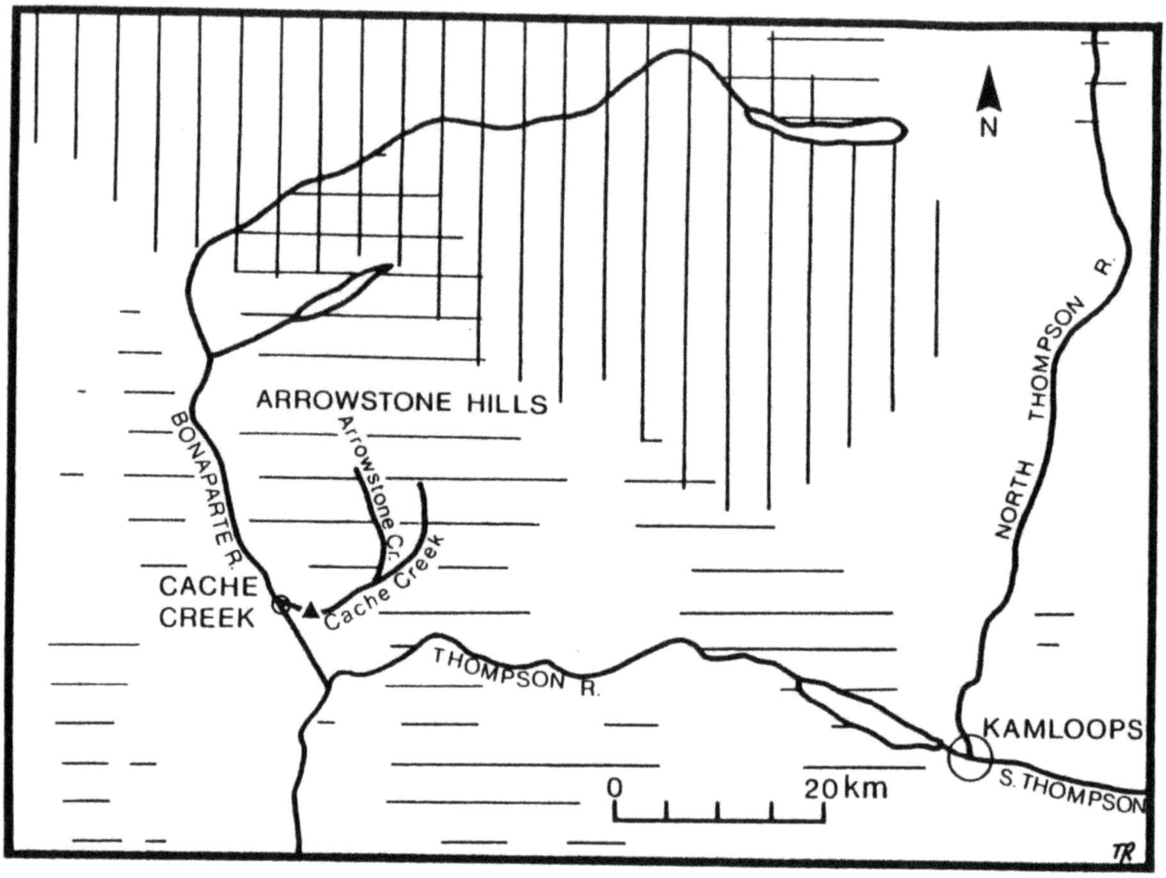

Figure 2. Location of the Cache Creek basalt source and the local distribution of Chilcotin Group and Kamloops Group basalts. Chilcotin Group basalts are indicated by vertical lines, Kamloops Group basalts by horizontal lines and the basalt collection area by a black triangle. Geological formation data from Bevier 1983a, Ewing 1981a, Duffel and McTaggart 1951, and Campbell and Tipper 1971.

Figure 3. Basalt cobble (arrow) in alluvial deposit along Cache Creek. Camera lens cap to right of cobble is 5 cm in diameter.

Figure 4. Basalt cobble (arrow) in till deposit exposed in bank along Cache Creek. Camera lens cap to right of cobble is 5 cm in diameter.

3.3 *Geological Origin and Petrography*

The basalt nodules collected for the present study came from the glacial and alluvial deposits and their bedrock source is not known at present. However, Ryder (1976:3) has observed that the local bedrock is strongly represented in the till, and while basaltic rocks going back to the Upper Triassic Nicola Group are present in the area (Duffell and McTaggart 1951:12-14), there are only two likely bedrock sources for the basalt in the till. Other potential sources of the basalt can be ruled out on the basis of their characteristics (e.g., colour, texture, composition, degree of weathering, etc.). The basalt was most likely derived from the Eocene "Kamloops Group" (Ewing 1981a,b) and/or the Miocene-Pliocene "Chilcotin Group" (Bevier 1982,1983a,b). Both of these formations are located near the surface, are abundantly represented in and around the Arrowstone Hills, and were likely to have been glacially scoured and included in till.

Originally defined by Dawson (1896), the "Kamloops Group" of south-central British Columbia has undergone redefinition and revision by subsequent researchers (e.g. Drysdale 1914; Duffel and McTaggart 1951; Campbell and Tipper 1971; Ewing 1981a,b). It consists of 1-3 km thick volcanic and sedimentary deposits dating from the Lower to Middle Eocene (ca. 42-52 million years)(Ewing 1981a,b). Several geologists have examined thin sections of Kamloops Group basalts (Duffell and McTaggart 1951:67; Ferrier 1896:384-386; Drysdale 1914:142). More recently, Ewing (1981b:1485) has studied thin sections of this material:

> "The most abundant rocks of the Kamloops Group volcanics are augite-pigeonite-labradorite andesites ranging in composition from 52 to 60% silica. Clots of clinopyroxene crystals (both augite and pigeonite), and stellate plagioclase aggregates are common... Euhedral olivine phenocrysts are observed in the more mafic samples (less than 55% silica)... Euhedral magnetite also forms phenocrysts in some samples. The groundmass typically consists of small plagioclase (andesine) laths with subhedral to anhedral interstitial augite and pigeonite, magnetite, and residual glass."

Campbell and Tipper (1971) distinguished late Miocene-early Pliocene "Plateau Lavas" from the Eocene "Kamloops Group" and Tipper (1978) later named them the "Chilcotin Group". Bevier (1982,1983a,b) studied this formation in detail and restricted the term to refer to volcanic materials, excluding sedimentary deposits included in Tipper's (1978) original definition. According to Bevier (1983a), the Chilcotin Group of lavas and associated breccias derive primarily from two periods of eruptive activity (ca. 2-3 million years ago and 6-10 million years ago) and form a flat lava plateau some 50,000 square kilometers in extent and at least 141 m thick in places. The Chilcotin

Group may also be the source for black, vitreous basalt found in the Baezaeko River area of the west-central interior of British Columbia.

Chilcotin Group basalt is described by Bevier (1983a:521-522) as consisting of

> "...up to 10% euhedral to anhedral olivine microphenocrysts in a subophitic to intersertal groundmass... Subophitic augite is present in most samples, and dityaxitic texture is common. Accessory minerals include ilmenite, magnetite, and apatite. In a few flows the olivine grains have altered partly or completely to iddingsite."

Elsewhere, Bevier (1983b:208) describes the basalt as containing

> "...olivine microphenocrysts in a ground-mass of titanaugite, plagioclase, and iron-titanium oxides... Groundmass texture ranges from subophitic to intersertal."

The basalt is phenocryst poor, with olivine usually being the only phenocryst present; the groundmass plagioclase is primarily labradorite and the iron-titanium oxides include both ilmenite and magnetite (Bevier 1983b:211-219). Petrographically, Chilcotin Group basalt is a typical alkali basalt, however, chemical analysis places it on the dividing line between alkaline and subalkaline and it therefore should be considered a transitional basalt (Bevier 1983b:209).

3.4 *Petrographic Analysis*

Thin sections of seventeen basalt cobbles were prepared and studied for two reasons: (1) to identify the geological source(s) of the basalt; (2) to identify specific characteristics which may be relevant to understanding differences in the microwear properties of vitreous and non-vitreous basalts (see Chapter 10.3).

Thin sections of basalt were prepared and studied by the author under the guidance of Dr. A. Mermut, Department of Soil Science, University of Saskatchewan, and his lab assistant, Glen Hudy. Samples were sawn into slabs approximately 1.5-2 cm long, 1-1.5 cm wide and .5 cm thick, using a Hillquist saw with a diamond cutting blade. Next, one face of the slab was ground smooth by using 25 µm grinding powder on a Logitech LP30 Production Lapping and Optical Polishing machine, and by hand grinding with 14 µm and lastly with 5 µm grinding powder in oil on a glass plate. Each slab was then mounted on a glass slide with Petropoxy 154 Resin and then cut to a thickness of 400 µm to 500 µm with the diamond saw. The slab was then ground to a thickness of 50 µm using 25 µm grinding powder on the Logitech machine and by hand to a thickness of 25 µm.

The thin sections were examined and photographed with a Zeiss "Universal R" binocular microscope with a M 35 photo-micrographic camera attachment. A Leitz Wetzlar monocular microscope was also used for preliminary examination. The minerals were mainly identified at magnifications of 100 x to 300 x and all photomicrographs reproduced here were taken at 156.3 x (Figures 5-14).

The pyroxene observed in the groundmass (Table 2) is most likely *augite* or *titanaugite* which has been observed in both Chilcotin Group basalts (Bevier 1983a:521, 1983b:208) and Kamloops Group basalts (Ewing 1981b:1485). *Pigeonite* was another type of pyroxene observed in Kamloops Group basalts (Ewing 1981b:1485). It is very likely that the groundmass "opaques" observed in the Cache Creek specimens are *ilmenite* and *magnetite* which have been identified in Chilcotin Group basalts (Bevier 1983b:211-219). The large, euhedral "opaques" are probably magnetite phenocrysts, which have been observed in some Kamloops Group basalt samples by Ewing (1981b:1485). Groundmass plagioclase has been observed in the Cache Creek samples and was identified as primarily *labradorite* in Chilcotin Group basalts (Bevier 1983b:208) and as *andesine* in Kamloops Group basalts (Ewing 1981b:1485).

Five of the cobbles were black in colour and the others ranged from medium grey to very dark grey. The black basalt samples were obviously fine-grained and even glassy in texture. This observation was confirmed by the microscopic analysis: the five black basalt samples had a very fine groundmass comprised of minute plagioclase laths, euhedral to anhedral olivine microphenocrysts, and rare occurrences of pyroxene and opaques, most of which consist of iron oxides (Figures 5-7). Phenocrysts are rare and they consist of euhedral to anhedral olivine crystals (Figure 8). The black basalt is most similar to the Chilcotin Group basalts as described by Bevier (1983a,b).

The grey-coloured basalt samples are coarser in texture than the black samples and usually contain significantly higher percentages of iron oxides than the black basalts. Large amounts of magnetite have only been noted in Kamloops Group samples (e.g. Ewing 1981b:1485; Drysdale 1914:142; Duffell and McTaggart 1951:67; Ferrier 1896:384-386), allowing the suggestion that most of this material originated from Kamloops Group deposits. Several grey samples also have feldspar phenocrysts (Figures 11-14), which constitute another characteristic of Kamloops Group basalts (Duffell and McTaggart 1951:67).

It is likely, then, that most grey basalt samples originated from the Kamloops Group and most black, fine-grained basalt samples from the more recent Chilcotin group. However, it is impossible to assign samples unequivocably to one or the other geological formations on the basis of presently available data due to the overlap in composition reported in published studies (Ewing 1981b; Bevier 1983b).

It is worth noting that the Chilcotin Group may also be the source for black, vitreous basalt found in the Baezaeko River area of the west-central interior of British Columbia. This basalt is visually indistinguishable from Cache Creek basalt and may well share the same petrographic and microwear characteristics.

Table 2. Cache Creek basalt thin section descriptions.

COBBLE	FIGURE	DESCRIPTION
22	5	Black, glassy, fine-grained basalt. Medium sized olivine phenocrysts in a very fine groundmass of plagioclase and rare opaques.
27	-	Black, glassy, fine-grained basalt. Occasional small olivine phenocrysts and clusters of olivine phenocrysts in a very fine groundmass of plagioclase, with rare opaques and pyroxene.
44	6	Black, glassy, fine-grained basalt. Occasional very large anhedral olivine phenocrysts in a very fine groundmass consisting of plagioclase and rare opaques.
21	7,8	Black, glassy, medium-fine grained basalt. Feldspar phenocrysts and large to very large anhedral olivine phenocrysts in a fine groundmass of plagioclase with rare opaques and pyroxene (augite).
117	-	Black, glassy, fine-grained basalt. Medium to very large euhedral to anhedral olivine phenocrysts in a very fine groundmass consisting of plagioclase.
11	-	Dark grey, fine-grained basalt. Occasional olivine microphenocrysts in fine-grained plagioclase groundmass also containing rare opaques and pyroxene.
23	-	Medium grey, fine-grained basalt. No phenocrysts are present in a fine textured plagioclase groundmass, containing rare opaques and pyroxene crystals. Extensive areas of yellowish-orange staining are usually associated with opaques.
40	-	Medium grey, medium basalt. Medium to large feldspar and olivine phenocrysts in a very fine-grained plagioclase groundmass, also containing opaques and pyroxene. Opaques are common and are associated with areas of yellow to dark yellow brown staining.
25	9	Medium greenish-grey, medium basalt. Medium to very large olivine phenocrysts in a medium coarse feldspar groundmass, also containing opaques and pyroxene. Yellow, orange and red staining is associated with the relatively common opaques and also cracks in the rock.
35	10	Very dark grey, medium basalt. Large to very large euhedral olivine phenocrysts are present in a medium coarse feldspar groundmass, also containing opaques and pyroxene. Yellow and orange staining associated with opaques.
2	-	Very dark grey, medium basalt. The occasional subhedral to euhedral olivine phenocryst and cluster of olivine phenocrysts are present in a medium coarse plagioclase groundmass, also containing opaques and pyroxene. Dark rusty-orange staining is associated with relatively common anhedral opaques.

continued

Table 2. (continued)

COBBLE	FIGURE	DESCRIPTION
20	-	Very dark grey medium-coarse basalt. Medium feldspar and large to very large olivine phenocrysts are present in a medium coarse groundmass of plagioclase, opaques and rare pyroxene. Orange and yellow staining are associated with the relatively common opaques. Occasionally, several olivine phenocrysts are grouped together forming a large mass.
31	-	Very dark grey, medium basalt. The medium-coarse groundmass of plagioclase, opaques and pyroxene contains occasional large to very large euhedral to subhedral olivine phenocrysts which are sometimes found in large clusters. Opaques are relatively small in size and commonly anhedral.
6	-	Very dark grey, coarse basalt. Large, euhedral feldspar phenocrysts are interspersed with subhedral to euhedral olivine phenocrysts and relatively large and common euhedral opaque phenocrysts.
8	-	Very dark grey, coarse basalt. Medium to large, euhedral feldspar phenocrysts are very common, as are subhedral olivine phenocrysts. Large opaque phenocrysts are common and are surrounded with orange brown staining.
26	11, 12	Very dark grey, coarse basalt. Feldspar phenocrysts are large, euhedral and randomly oriented. Medium to large subhedral to euhedral olivine phenocrysts are commonly interspersed between the feldspars. These appear to have been altered (iddingsitization?). Unaltered euhedral opaque phenocrysts are relatively large and common.
34	13, 14	Very dark grey, very coarse basalt. Euhedral feldspar phenocrysts are often very large, although large and medium ones are most common. Subhedral to euhedral olivine phenocrysts vary from small to very large, and appear to be partly weathered to iddingsite. Opaques phenocrysts are small to large, anhedral to euhedral and very common.

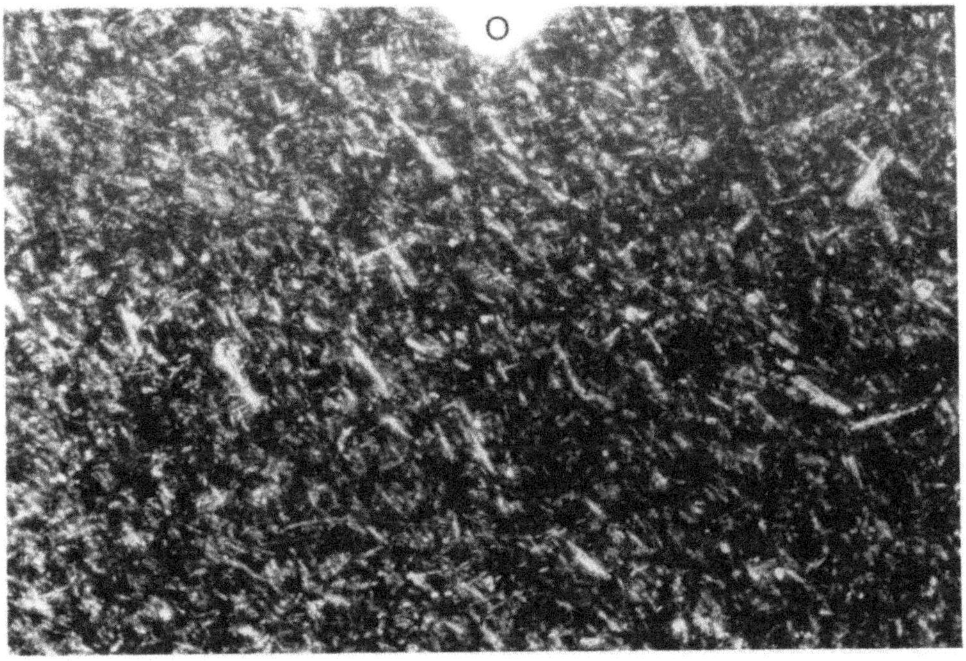

Figure 5. Thin section of cobble 22, black, fine-grained basalt. Very fine plagioclase groundmass with one olivine (O) phenocryst. Polarized light, 546 x.

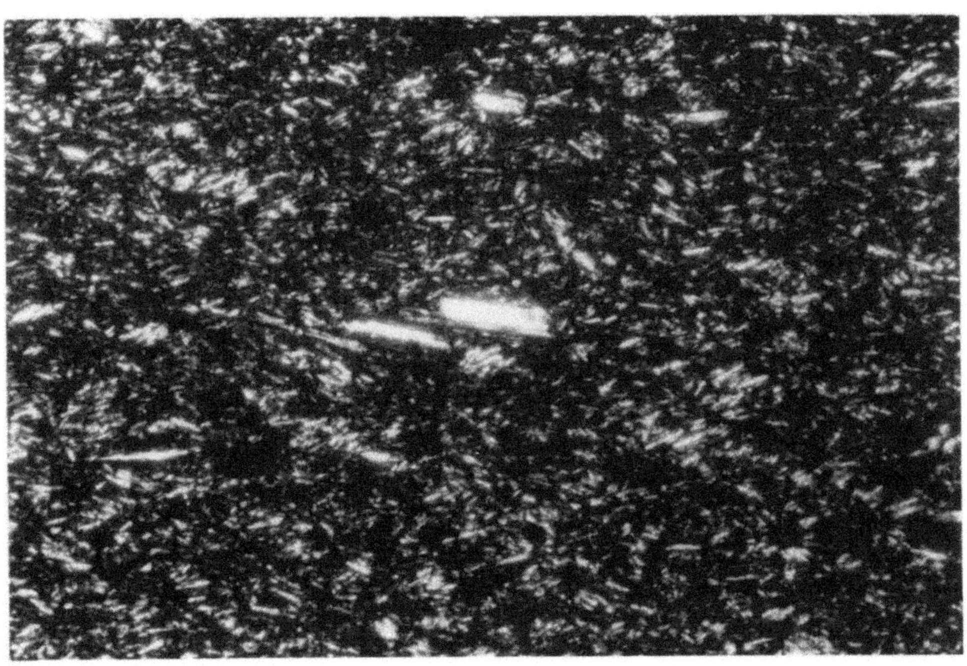

Figure 6. Thin section of cobble 44, black, fine-grained basalt. Very fine plagioclase groundmass. Polarized light, 546 x.

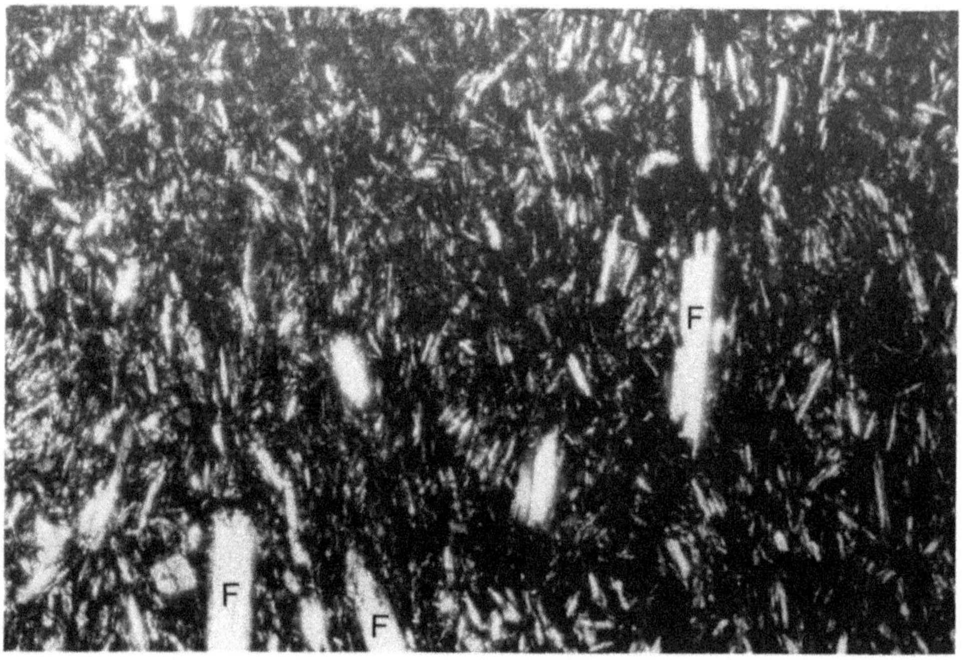

Figure 7. Thin section of cobble 21, black, medium-fine grained basalt. Fine plagioclase groundmass with feldspar (F) phenocrysts. Polarized light, 546 x.

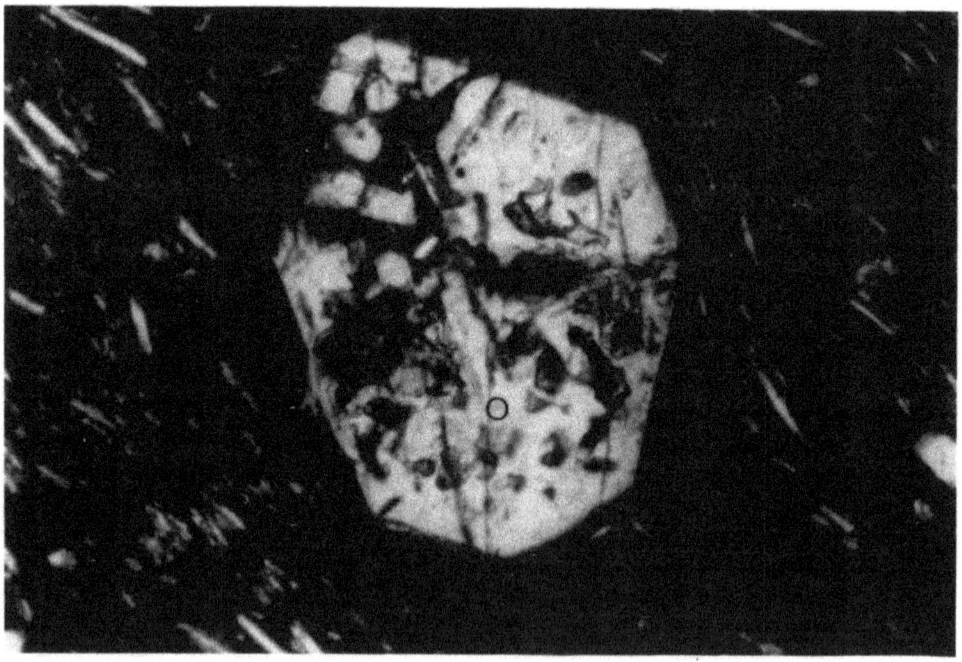

Figure 8. Thin section of cobble 21, new area. Olivine (O) phenocryst in plagioclase groundmass. Polarized light, 546 x.

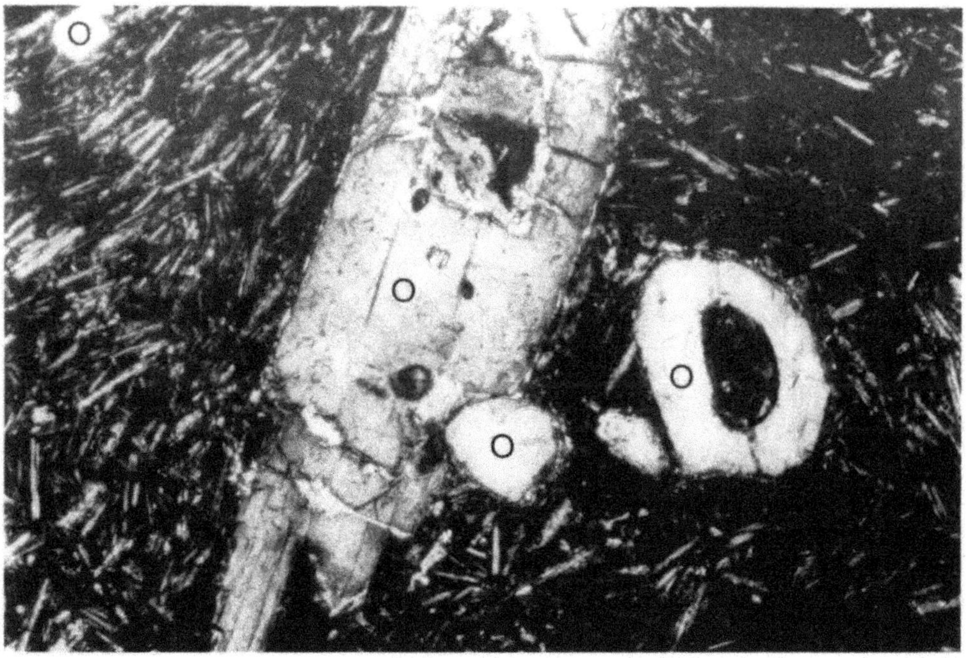

Figure 9. Thin section of cobble 25, greenish-grey, medium grained basalt. Medium-coarse plagioclase groundmass and olivine (O) phenocrysts. Polarized light, 546 x.

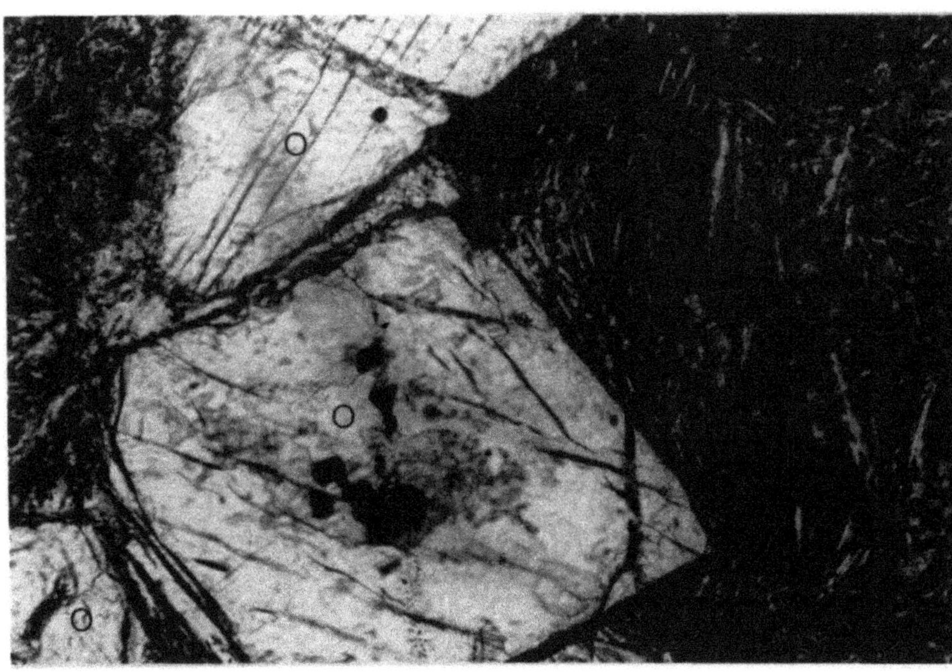

Figure 10. Thin section of cobble 35, very dark grey, medium-grained basalt. Medium-coarse plagioclase groundmass and olivine (O) phenocrysts. Polarized light, 546 x.

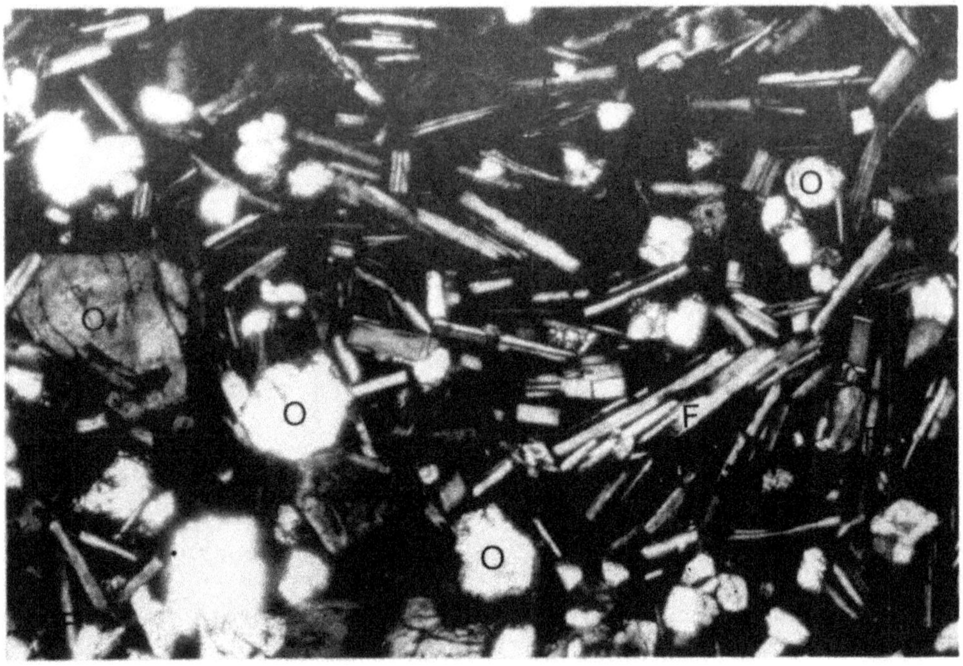

Figure 11. Thin section of cobble 26, very dark grey, coarse-grained basalt. Feldspar (F) and olivine (O) phenocrysts. Polarized light, 546 x.

Figure 12. Thin section of cobble 26. Feldspar (F), olivine (O), and opaque (Op) (magnetite) phenocrysts. Same area as Figure 11 under normal light, 546 x.

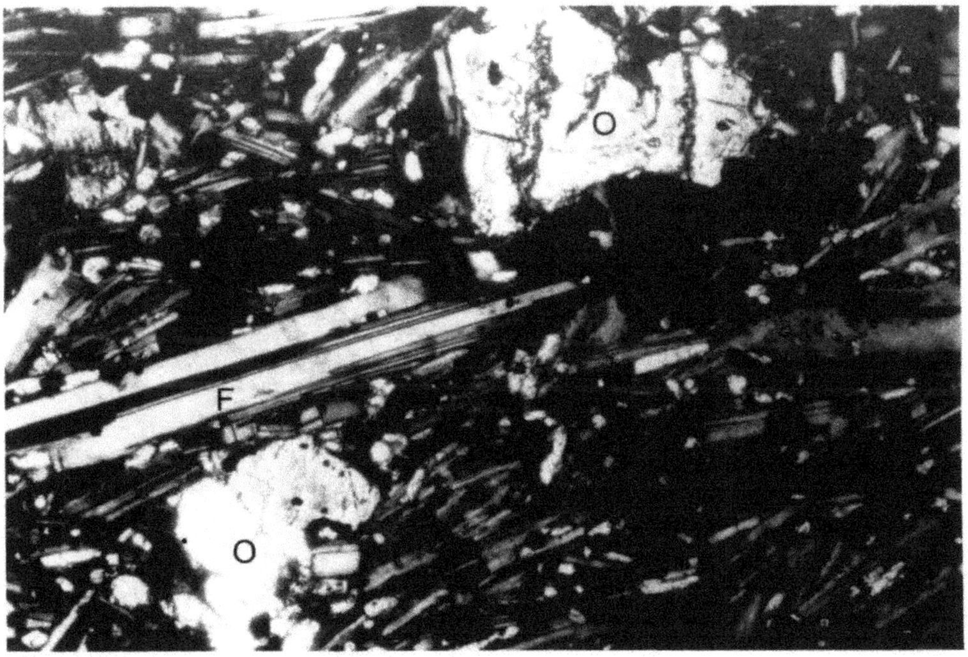

Figure 13. Thin section of cobble 34, very dark grey, very coarse basalt. Feldspar (F), opaque (Op) (magnetite), and weathered olivine (O) phenocrysts. Polarized light, 546 x.

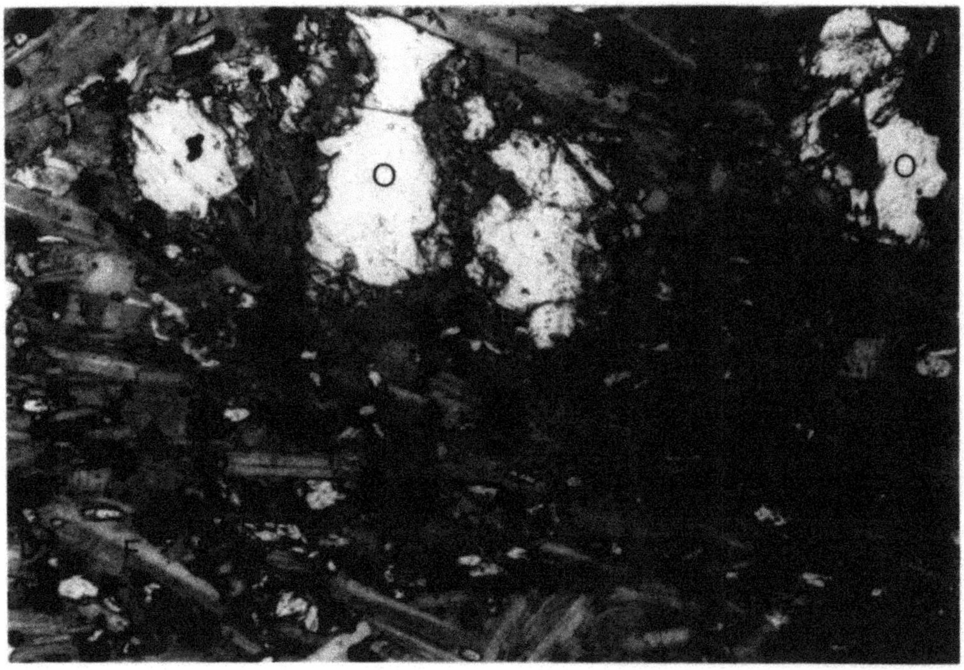

Figure 14. Thin section of cobble 34. Feldspar (F), opaque (Op) (magnetite), and weathered olivine (O) phenocrysts. Normal light, 546 x.

CHAPTER 4

HISTORIC AND PREHISTORIC USE OF CACHE CREEK BASALT

4.1 *Introduction*

Historically, the southwestern interior of British Columbia was occupied by the Interior Salish speaking Shuswap, Thompson and Lillooet (Fraser Canyon Lillooet) people (Figure 15). They were egalitarian, semi-sedentary, hunter-gatherers, largely dependent on salmon for subsistence. Permanent pithouse villages were occupied during the winter, but populations moved to temporary field camps and stations to exploit spatially segregated, seasonally-available resources during the rest of the year (Teit 1900,1906,1909; Boas 1891; Dawson 1891; Ray 1939). A similar way of life has been documented by archaeologists for at least the past 3500-4000 years in the territory occupied by the Shuswap, Thompson and Lillooet (Sanger 1970; Stryd 1973a,b; Wilson 1980; Carlson 1980; Wyatt 1972; Rousseau and Richards 1985), and is now known as the *Plateau Pithouse tradition* (Richards and Rousseau 1982,1983,1987).

Historically and prehistorically, basalt was the most common lithic material used for making chipped stone tools in this part of British Columbia. This chapter documents the technology used to make chipped stone tools in historic and prehistoric times in the area, along with descriptions of the kinds of tools which were made, how they were used and what they were used on. This information is used to provide a systemic framework for the basalt tool-use experiments described in Chapter 7.

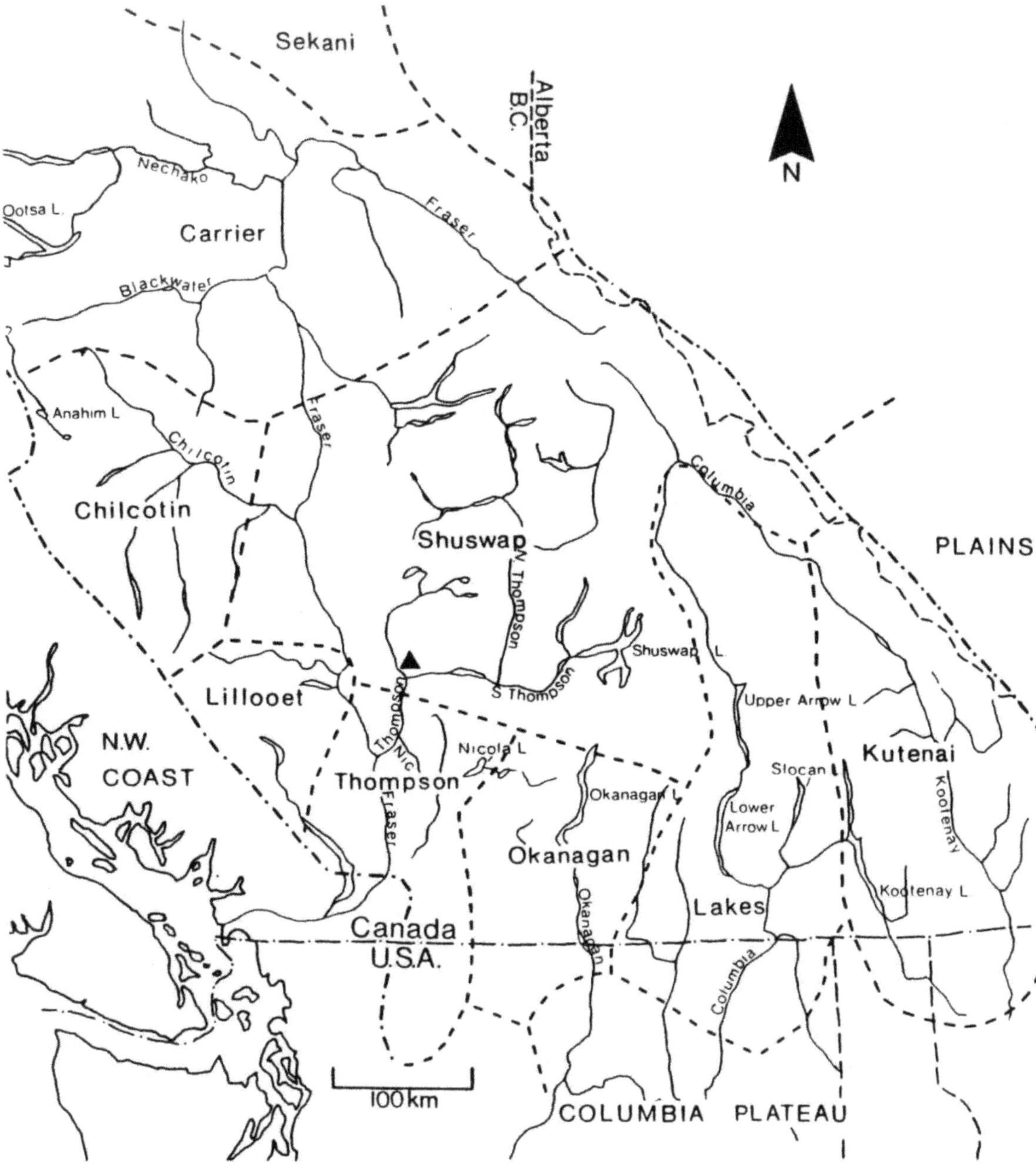

Figure 15. Ethnographic map of the southern interior of British Columbia showing mid-nineteenth century tribal locations (after Duff 1964). The black triangle indicates the location of the Cache Creek basalt collection area used for this study.

4.2 *Thompson, Shuswap and Lillooet Lithic Technology*

James Teit (1900,1906,1909), principal ethnographer of the Interior Salish, had observed that making tools by chipping stone was rare by the beginning of the 20th century. However, both Teit (1900:182) and Smith (1899:136) found Thompson Indians living near Lytton who still knew how to manufacture chipped stone arrow points. Chipped-stone hide scrapers were documented as still being in use in the late 19th and early 20th centuries (Boas 1891:636; Smith 1899:147; Teit 1900:182). Direct observation of chipped stone tool manufacture and use was thus possible for turn of the century ethnographers. Even in the mid-1930s, elderly informants were still able to provide some information about chipped stone tools (Ray 1942).

"Glassy basalt" was the most common lithic material used for making chipped stone tools, although jasper, obsidian, quartz, quartzite, "flint", chalcedony, and agate were also occasionally used (Dawson 1891:35; Teit 1900:241, 1906:203,225,234, 1909:473; Ray 1942:147,152; Smith 1899:135-137). Dawson (1891,1896) studied both the geology and ethnology of southwestern interior British Columbia in the late 19th century and consequently was well acquainted with the basalt lithic material, its aboriginal use, and its source. He observed that the basalt was

> "...very common in loose boulders and fragments everywhere in the vicinity of the Arrowstone Hills. It was formerly employed by the Indians for the manufacture of arrow- and spear-heads, and probably occurs in considerable volume in the higher parts of the Arrowstone Hills" (Dawson 1896:220).

It appears the basalt occurred abundantly as cobbles over a wide area, and Dawson seems to suggest a bedrock source in the higher elevations of the Arrowstone Hills. The existence of a bedrock quarry used by the Thompson Indians was recorded by Smith, although he only vaguely notes the location as "the mountains" (Smith 1899:137). The general source of the basalt was confirmed by Teit. According to his informants, the Thompson obtained basalt "...at a certain place north of Thompson River" (Teit 1900:241), and his Shuswap informants stated it was found "...in many parts of the country, notably in the Arrow-stone Range, near Bonaparte River" (Teit 1909:519). Basalt used to make tools for the present study came from Cache Creek, which is north of the Thompson River, just east of the Bonaparte River and on the fringes of the Arrowstone Hills (see Chapter 3.1, Figures 1 and 2).

The source of basalt used by the Lillooet was not recorded, although it is probably significant that the boundary between the three groups is close to the Arrowstone Hills, making the source accessible for direct procurement by members of each group. If it could not be obtained by direct procurement, there was considerable interaction between the three groups which would have facilitated the circulation of this material.

Teit describes a method of chipped stone arrowhead manufacture used by the Thompson Indians as follows:

> "The bowlder was split by being laid on a stone and struck with a hand-hammer, generally a pebble of handy size. When a suitable piece had been obtained, its edges were trimmed off with a hard stone. Then it was wrapped in grass or hay, placed on edge on a stone, and large flakes were split off with a hand-hammer. After a suitable piece had been obtained, it was placed on a pad in the left hand and held in position with the fingers. It was given its final shape by means of a flaker made of antler, which was used with a forward and downward pressure. The blunt point served for flaking off larger chips, while the smaller one was used for the final stages of the work" (Teit 1900:182, Figure 118).

It is apparent that the Thompson used bipolar percussion, direct freehand percussion with hard hammers, and pressure flaking with antler and bone flakers to manufacture chipped stone points and probably other types of tools. According to Teit (1906:203; 1909:473), Lillooet and Shuswap lithic technology was virtually identical to that of the Thompson.

Arrow points were side-notched, barbed or leaf-shaped in form; spearheads, axe-heads ("tomahawks") and daggers were of similar forms but larger (Teit 1900:241, 1906:234). In addition to the above chipped stone tool types, knives, scrapers and drills were also used (Teit 1900:182-184,263, 1906:203, 1909:473,474; Ray 1942:126, 146,147,150,152,162; Boas 1891:636,638; Dawson 1891:17; Smith 1899:147).

While the exact method of manufacture of stone knives is not recorded in the ethnographies, they are described as having been made by being "chipped and flaked" (Teit 1909:473), "worked and flaked" (Teit 1906:203), and "battered into shape, cut and flaked" (Teit 1900:182). Of the two knives illustrated by Teit (1900:Figures 125,126), one is extensively chipped, while the other has only marginal unifacial retouch along most of the edge and bifacial retouch on its base. Chipped-stone knives were used for cutting or whittling wood, hide, antler and bone (Teit 1900:183,187; Teit 1909:474,508; Ray 1942:147), butchering fish (Ray 1942:126), and were presumably used on other materials which needed to be processed or manufactured by slicing, sawing or whittling (see Tables 3-7). Knives and daggers used in warfare were often hafted in short wooden or antler handles (Teit 1906:234; Ray 1942:152-153). Other types of knives were hafted in bone, antler and wooden handles and bound into place with sinew, or were unhafted but wrapped in buckskin (Teit 1900:183, 1909:474; Ray 1942:126,147).

Hide scrapers of stone apparently continued to be in use into the 20th century. The Thompson hide defleshing tool is described only as a "sharp stone scraper" (Teit 1900:184). The Lillooet hide scrapers "...were simply thin pieces flaked off from pebbles of various kinds, and were slightly chipped on one edge only" (Teit 1906:203). Among the Shuswap, hide scrapers "...were usually made from chips of bowlders which split well, but occasionally were flaked from basalt, in the same manner as the arrow-points" (Teit 1909:473). It appears that basalt was only occasionally used for making chipped stone hide scrapers while other materials such as schist were more commonly used (Ray 1942:126). In general, hide scrapers seem to have been minimally retouched, and probably only unifacially. An illustrated example of a hide scraper (Smith 1899:Figure 64) is attached to a 1 m long wooden handle by rawhide binding. The scraper had marginal retouch around the entire illustrated face. In addition to hide scraping, chipped stone tools were also used as scrapers to straighten arrow shafts (Ray 1942:150). There can be little doubt that chipped stone scrapers were used on many materials which needed to be manufactured or processed by scraping or shaving.

Bits made of basalt or "flint" were used for drilling a variety of materials, including steatite (Teit 1900:183, 1909:474; Ray 1942:146). Drill bits were apparently made on "flaked...points" (Teit 1909:474), some of which were "notched" (Ray 1942:146), which suggests that at least some stone drill bits were made of chipped stone projectile points with narrowed tips. Projectile points with such modifications have, in fact, been occasionally found at prehistoric sites in the area (e.g. Wilson 1980: Figure 45e-i). Stone drill bits were fitted to shafts which were hand-rotated (Teit 1909:474; Ray 1942:146).

The Thompson, Shuswap and Lillooet had a division of labour by sex, with the manufacturing of stone tools recorded as being a male task (Teit 1900:182,295). Indeed, all "work in stone, bone and wood was done by the men" (Teit 1900:182). However, women did engage in activities which would have involved the *use* of stone tools: the preparation of hides (scrapers); the manufacture of matting, basketry, twine and clothing (knives); the processing of fish (knives); and the preparation of meals (knives) (Teit 1900:182,230,296; Ray 1942:124; Smith 1899:147; Bouchard and Kennedy 1977:66, 1979:134,146). It is difficult not to speculate that women occasionally manufactured or at least resharpened their own stone scrapers and knives, even though there is no documented evidence of such behaviour.

In the ethnographic record reviewed above, chipped stone tools were used with the following motor actions: scraping, cutting (sawing, slicing), whittling ("carving"), drilling, chopping ("tomahawks"), and piercing (daggers). Also, arrow and spear heads were used as projectile tips. Engraving tools would have been useful for producing the incised designs found on bone, antler and stone implements, although gravers are not mentioned in the ethnographies. Perhaps such implements were replaced by much more efficient metal tools at such an early date that they were forgotten by the late 19th century.

It is apparent that chipped stone tools were formed by variable degrees of retouch, with projectile points undergoing the most elaborate manufacturing process. In contrast, some hide scrapers were not retouched and some knives were only minimally retouched. Other simple unretouched or minimally retouched flake tools were probably also used, although no mention of such tools is made in the ethnographies. The author speculates that simple flake tools may have easily been overlooked by informants or ethnographers.

In summary, chipped stone tools, primarily made of basalt, were used by both men and women on a variety of contact materials. These include animals which were processed for subsistence purposes (Tables 3,4), plants and trees used for technological purposes (Table 5), and a variety of materials which were modified into weapons, tools, containers, clothing, ornaments, or gaming pieces (Table 6).

4.3 *Prehistoric Use of Basalt*

Lithic assemblages from archaeological sites in Shuswap, Lillooet and Thompson territory usually consist of high percentages of items made of basalt (see Sanger 1968a,1970; Richards 1978; Richards and Rousseau 1982,1987; Rousseau and Richards 1985,1988; Stryd and Lawhead 1983; Stryd 1972,1973a,1974,1981; Eldridge and Stryd 1983; Pokotylo and Beirne 1978; Whitlam 1980; Wyatt 1972; Von Krogh 1978; Wilson 1980; Carlson 1980). This basalt is visually indistinguishable from basalt collected at Cache Creek for the present study, or basalt widely available in glacial, fluvio-glacial or fluvial deposits in the area. Basalt frequently comprises 90% or more of chipped stone assemblages at prehistoric sites in the study area (see above references). In addition, earlier assemblages tend to have high proportions of non-vitreous basalts, while vitreous basalts often comprise approximately 80% of those assemblages dating more recently than ca. 2500 B.P. (Richards and Rousseau 1987:27,34).

There are some chipped stone tool types found at *Plateau Pithouse tradition* (*PPt*) sites which are not described or illustrated in the ethnographies; similarly, some tools described in the ethnographies have not been found at archaeological sites (Table 7). Tools described by archaeologists as "gravers", "choppers" "split cobble tools" and "microblades" are not described in the ethnographies, but are found in *PPt* sites, although none of these tool types are common. Chipped stone "daggers" and "axeblades" are mentioned in the Thompson ethnography by Teit (1900) but such items have not been identified in prehistoric assemblages. In most *PPt* assemblages from historic Shuswap, Thompson and Lillooet territory, minimally retouched and unretouched tools ("utilized flakes") predominate lithic tool assemblages, while the well-made "projectile points", "knives" and "scrapers" are generally less common.

The Shuswap, Lillooet and Thompson manufactured tools using bipolar percussion, direct freehand percussion with hard hammers, and pressure flaking using antler tine flakers or bone tips in antler tine holders. All of these techniques have been identified on *PPt* tools or cores (Richards and Rousseau 1987). In addition, soft hammer percussion may have been used for bifacial thinning (Magne and Pokotylo 1981), although billets have not yet been found. Microblade technology was present until approximately 1200 B.P. (Sanger 1968b,1970; Richards and Rousseau 1987).

Table 3. List of mammal species exploited historically by the Shuswap, Lillooet and Thompson Indians (Teit 1900,1906,1909; Ray 1942) and prehistorically in the Plateau Pithouse tradition (Richards and Rousseau 1982,1983,1987).

COMMON NAME	SCIENTIFIC NAME	SHUSWAP LILLOOET THOMPSON	PLATEAU PITHOUSE TRADITION
mule deer	*Odocoileus hemionus hemionus*	+	+
black-tailed deer	*Odocoileus hemionus columbianus*	+	
white-tailed deer	*Ococoileus virginianus ochrourus*		+
elk	*Cervus canadensis*	+	+
bighorn sheep	*Ovis canadensis*	+	+
mountain goat	*Oreamos americanus*	+	
mountain caribou	*Rangifer tarandus*	+	+
black bear	*Ursus americanus*	+	+
grizzly bear	*Ursus horribilis*	+	
cougar	*Felis concolor*	+	+
lynx	*Lynx canadensis*	+	
wolverine	*Gulo luscus*	+	
coyote	*Canis latrans*	+	+
wolf	*Canis lupus*	+	+
fox	*Vulpes spp.*	+	
red fox	*Vulpes fulva*		+
dog	*Canis familiaris*	+	+
weasel	*Mustela spp.*	+	
marten	*Martes americana*	+	+
mink	*Mustela vison*	+	
fisher	*Martes pennati*	+	
otter	*Lutra canadensis*	+	
muskrat	*Ondatra zibethicus*	+	+
beaver	*Castor canadensis*	+	+
marmot	*Marmota spp.*	+	+
porcupine	*Erethizon dorsatum*	+	+
snowshoe hare	*Lepus americanus*	+	+
rabbit	*Sylvilagus spp.*	+	+
western grey squirrel	*Sciurus griseus*	+	+
red squirrel	*Tamiasciurus hudsonicus*		+
striped skunk	*Mephitis mephitis*		+
packrat	*Neotoma sp.*		+

Table 4. List of fish species exploited historically by the Shuswap, Lillooet and Thompson Indians (after Teit 1900,1906,1909; Ray 1942; Kennedy and Bouchard 1975,1978) and prehistorically in the Plateau Pithouse tradition (Richards and Rousseau 1982,1983,1987).

COMMON NAME	SCIENTIFIC NAME	SHUSWAP LILLOOET THOMPSON	PLATEAU PITHOUSE TRADITION
spring salmon (chinook)	*Oncorhynchus tshawytscha*	+	+
pink salmon (humpback)	*O. gorbuscha*	+	+
coho salmon	*O. kisutch*	+	
sockeye salmon	*O. nerka*	+	+
kamloops or rainbow trout (steelhead)	*Salmo gairdneri*	+	+
Dolly Varden char	*Salvelinus malma*	+	
burbot	*Lota lota*	+	
sturgeon	*Acipenser spp.*	+	
whitefish	*Prosopium spp.*	+	
largescale sucker	*Catostomus macrocheilus*	+	
suckers	*Catostomus spp.*	+	
eels (Pacific lamprey)	*Entosphenus tridentatus*	+	
peamouth	*Mylocheilus caurinus*	+	
squawfish	*Ptychocheilus oregonensis*	+	
redside shiner	*Richardonius balteatus*	+	
sculpin	*Cottus spp.*	+	

Table 5. Plant and tree species used for technological purposes in historic times by the Shuswap, Lillooet and Thompson (Teit 1900,1906,1909; Ray 1942; Turner 1979) and which may have been processed or modified with chipped stone tools.

COMMON NAME	SCIENTIFIC NAME
horsetails	*Equisetum spp.*
common horsetail	*E. arvense*
common scouring rush	*E. hiemale*
sword-fern	*Polystichum munitum*
bracken fern	*Pteridium aquilinum*
yellow cedar	*Chamaecyparis nootkatensis*
Rocky Mountain juniper	*Juniperus scopulorum*
western red cedar	*Thuja plicata*
grand fir	*Abies grandis*
amabilis fir	*Abies amabilis*
subalpine fir	*Abies lasiocarpa*
engelman spruce	*Picea engelmannii*
white spruce	*Picea glauca*
lodgepole pine	*Pinus contorta*
white pine	*Pinus monticola*
ponderosa pine	*Pinus ponderosa*
douglas fir	*Pseudotsuga menziesii*
western yew	*Taxus brevifolia*
sedges	*Carex spp.*
tule	*Scirpus acutus*
grasses	*Poaceae or Gramineae*
pinegrass	*Calamagrostis rubescens*
ryegrass	*Elymus cinereus*
reed canary-grass	*Phalaris arundinacea*
common reedgrass	*Phragmites communis*
cat-tail	*Typha latifolia*
rocky mountain maple	*Acer glabrum*
cow parsnip	*Heracleum lanatum*
Indian hemp	*Apocynum cannabinum*
spreading dogbane	*Apocynum androsaemifolium*
showy milkweed	*Asclepias speciosa*
pasture wormwood	*Artemisia frigida*
big sagebrush	*Artemisia tridentata*
balsamroot	*Balsamorhiza sagittata*
rabbitbrush	*Chrysothamnus nauseosus*

continued

Table 5. (continued)

COMMON NAME	SCIENTIFIC NAME
red alder	*Alnus rubra*
western white birch	*Betula papyrifera*
hazelnut	*Corylus cornuta*
prickly-pear cactus	*Opuntia fragilis* or *polyacantha*
orange honeysuckle	*Lonicera ciliosa*
red elderberry	*Sambucus racemosa*
blue elderberry	*Sambucus cerulea*
waxberry	*Symphoricarpos albus*
flowering dogwood	*Cornus nuttallii*
red-osier dogwood	*Cornus stolonifera*
silverberry	*Elaeagnus commutata*
wild gooseberries	*Ribes divaricatum*
mock-orange	*Philadelphus lewisii*
white clematis	*Clematis ligusticifolia*
saskatoon	*Amelanchier alnifolia*
black hawthorn	*Crataegus douglasii*
oceanspray (ironwood)	*Holodiscus discolor*
bitter cherry	*Prunus emarginata*
wild rose	*Rose acicularis*
thimbleberry	*Rubus parviflorus*
black cottonwood	*Populus balsamifera*
trembling aspen	*Populus tremuloides*
willow	*Salix spp.*

Table 6. List of organic (wood, bone, antler, shell, tooth, claw and hide) material culture items partly or completely manufactured with chipped stone basalt tools by the historic Shuswap, Lillooet and Thompson Indians (Teit 1900,1906,1909; Ray 1942) and in the Plateau Pithouse tradition (Richards and Rousseau 1982,1983,1987).

ITEM	SHUSWAP LILLOOET THOMPSON	PLATEAU PITHOUSE TRADITION
arrow foreshaft	b,a	
arrow shaft	w	
spear shaft	w	
harpoon shaft	w	
fire drill shaft	w	
bag net handle	w	
scraper handle	w	
fish gaff handle	w	
elbow adze handle	w,b	
chisel handle	w,a	
dagger handle	w	
knife handle	w,b,a	
tomahawk handle	w	
digging stick handle	w,a	
chisel sleeve	a	
adze sleeve	a	
sleeve or handle	a	
bag net hoop	w	
bag net ring	a	
point		b,a
spear point	b	
barbed point	w,b,a	b,a
composite harpoon valve	w,b,a	a,b
armour	w,h	
sheild	w,h	
bow	w	
mineature bow	b	
club	w,b,a	a
pestle	w	
mallet	w,b,a	
wedge	w,a	a,b
chisel	b,a	
dagger	b	
knife	b,a	
paddle	w	
pressure flaker	a	a

continued

Table 6. (continued)

ITEM	SHUSWAP LILLOOET THOMPSON	PLATEAU PITHOUSE TRADITION
drill bit	b,a	
digging stick	w,b	
gorget	b	
fish hooks	w,b,a	
fish spreaders	w	
stirrer	w	
spoon	w,b,a	
spatula	w	
reamer	b	
mat creaser		b
hide scraper	w,b	
bark peeler	w,a	
cambium scraper	b,a	a
tongs	w	
pin		b
needle	w,b,a	b
awl	w,b,a	b
netting stick	w	
hair tweezers	w,a	
ring		b
whistle	b	
scratcher	b	
drinking tube	b	b
bead	b,a,t,s	b,t,s,c
nose ornament	b,s	
bracelet		b
pendant		b,t,s,c
comb	w	a
anthropomorphic figurine		a
carved bear baculum		b
gambling sticks	w	
dice	t	t
bow hand guard	h	
quiver	h	
bag	h	
head band	h	
cap	h	
clothing	h	
moccasins	h	
mittens	h	

w=wood, b=bone, a=antler, h=hide, s=shell, t=tooth, c=claw

Table 7. List of chipped stone tool types used historically by the Shuswap, Lillooet and Thompson (Teit 1900,1906,1909) and prehistorically in the Plateau Pithouse tradition (Richards and Rousseau 1982,1983,1987).

SHUSWAP/LILLOOET/THOMPSON	PLATEAU PITHOUSE TRADITION
arrow points	arrow points (formed bifaces)
spear points	spear points (formed bifaces)
knives	knives (formed unifaces or bifaces, unformed unifaces or bifaces, utilized flakes)
daggers	no equivalent (possibly formed bifaces)
scrapers	scrapers (formed unifaces, unformed unifaces, spall tools, split cobble tools, utilized flakes)
notched drills	notched drills (formed bifaces, reworked points)
drills	drills (formed bifaces)
axeblades	no equivalent
no equivalent	perforator/gravers

4.4 *Summary and Discussion*

The historic Shuswap, Lillooet and Thompson Indians and their prehistoric ancestors of the *Plateau Pithouse tradition* were heavily reliant on chipped basalt tools for obtaining and processing subsistence resources and for manufacturing many other implements important for undertaking subsistence related tasks. The glassy basalt, which was preferred for tool making due to its superior flaking characteristics, was widely available as float cobbles but only concentrated in a few localities such as the Arrowstone Hills. The basalt used in the present study was obtained from Cache Creek on the periphery of the Arrowstone Hills.

The tools used in the present study are primarily simple flake tools (Appendix D), which are the most common implements found at *PPt* sites (Richards and Rousseau 1987). In the *PPt* and during the historic period, tools were manufactured mainly by hard hammer percussion and also by soft hammer percussion; such is also the case with the experimental tools used in this study. Tool actions used historically and in the *PPt* include cutting, scraping, whittling, drilling, boring, graving, piercing (daggers, projectile points), adzing and chopping. Ground nephrite celts were used for adzing and chiseling. Chipped pebble or cobble tools, usually not made of basalt, were used for chopping. With the exceptions of drilling (which is a very similar action to boring) piercing and chopping, all of the recognized aboriginal tool actions involving chipped stone tools were used in the experiments. The wide variety of plant and animal species exploited historically and prehistorically by Native people for subsistence purposes and for raw materials are listed on Tables 3-5. The scope of the present study only allowed the use of a limited range of contact materials, although those selected are ones which were commonly used and are representative of the major classes of contact materials.

CHAPTER 5

INDEPENDENT VARIABLES, RECORDING METHODS, AND PROCEDURES

5.1 *Introduction*

Basalt tools were used in experiments to determine the effects of different use-actions and contact materials on the tool edges. Characteristics of the basalt tools and the contact materials, and how the tools were used on the contact materials are the *independent* variables in the experiments, while microwear traces on the tools resulting from use are the *dependent* variables (see Myers 1972:1). Characteristics of the independent variables are the subject of this chapter, while the dependent variables are discussed in Chapter 6.

Independent variables involved in the experiments can be divided into three groups. Qualities of the tools comprise one group and include such attributes as type of basalt, spine-plane angle and various other characteristics of the working edge (Table 8). Another group of independent variables are attributes of the contact material, including hardness, elasticity, wetness, grittiness, etc. The last group of variables consists of observations relating a tool to a contact material during the active part of an experiment. Included in this group are tool action, applied pressure, angle of the tool edge to the contact surface, and duration of use. The above variables are defined and described in sections 5.2-5.3 and the actual states of the independent variables involved in each experiment are presented in tabular form on Table 9.

As discussed briefly in Chapter 2.1, other microwear researchers characterize experiments as "mechanical" or "goal oriented". The experiments carried out for the present study lean more toward "goal oriented" than "mechanical" in

Table 8. Experimental tool pre-use attributes. Tools are grouped by contact materials. A key to the abbreviations used is at the end of the table.

TOOL NO.	L	W	T	B	SPINE ANGLE	EOS	EC	R	D
PLANTS									
12-9	4.9	2.0	.9	v	25-40	i	p	n	y
12-12	5.4	5.7	1.5	v	20-45	i	p	n	y
134-11	6.8	5.2	1.1	v	20-25	i	p	n	n
43-2	2.5	3.5	.7	v	20-35	s	p	n	n
161-11	3.6	3.4	1.0	v	25-40	i	s	n	n
HIDE									
1-3	3.2	3.3	1.0	n	35-55	v,i	s	n	n
12-13	4.4	4.1	.8	v	50	v	s	n	y
17-14	5.6	6.9	2.1	v	50-80	v	p	n	n
22-7	3.0	2.6	.9	v	50	v	p	n	y
23-3	4.5	2.7	1.2	n	50-60	v,i	p	n	n
27-16	4.1	5.3	1.2	v	30-40	v,c	p	n	y
39-1	3.0	2.9	.9	v	40-50	v	s	n	y
44-4	3.3	2.6	.5	v	40-50	v,i	p	n	y
3-3	3.6	4.0	1.1	n	60	c	s	n	n
3-4	3.4	3.3	.7	n	60	s	p	n	n
22-12	4.3	3.0	1.2	v	50	s,v	s	n	y
27-17	4.4	3.3	1.6	v	50-60	s,v	s	n	y
31-2	2.6	2.9	.9	n	45-50	v,i	p	n	n
44-6	4.4	3.2	1.8	v	50-90	c,i	s	n	y
46-4	3.6	3.6	.8	v	30-40	i	p	n	n
80-4	6.0	3.9	2.0	v	70-80	v	p	u	-
80-5	5.5	3.6	1.9	v	60-70	v,i	p	u	-
80-6	4.9	6.4	1.7	v	30	i	n	n	-
80-7	3.9	2.9	.9	v	30-40	i	p	n	-
80-8	3.5	3.8	.6	v	20-30	s,i	s	n	-
12-17	3.4	4.2	.9	v	20-50	s	s	n	n
134-9	6.3	5.2	1.1	v	40-50	i	p	n	n

continued

Table 8. (continued)

TOOL NO.	L	W	T	B	SPINE ANGLE	EOS	EC	R	D
BONE									
21-6	2.6	3.6	.9	v	45-50	s	n	n	y
26-8	3.8	4.1	.9	n	30	s	s	n	n
26-9	4.2	3.9	.7	n	10-30	c	p	n	n
46-2	3.4	3.5	.9	v	30-40	i	s	n	n
11-6	3.3	3.0	.8	n	35-40	v	p	n	n
18-19	6.4	2.8	1.3	v	40-50	i	p	n	n
22-20	4.0	2.2	.9	v	30-50	s,i	s	n	y
17-13	6.2	4.4	1.2	v	60	c	s	n	n
63-1	3.7	4.3	1.0	v	30-40	i	s	n	n
138-19	3.9	4.9	1.2	v	-	-	-	n	n
ANTLER									
43-3	3.2	3.3	.7	v	30-40	v	p	n	y
134-22	9.0	4.1	1.8	v	30-50	i	s	n	n
141-7	3.6	1.9	.5	v	25-40	i	p	n	y
11-2	4.8	4.1	1.4	n	40-50	s,i	p	n	n
11-12	6.8	5.1	2.0	n	60	c	p	n	n
17-11	3.4	3.7	.7	v	20	i	p	n	n
27-3	2.2	2.9	.4	v	40-50	c	s	n	n
36-7	4.1	1.3	.4	v	20	i	s,p	n	y
18-17	3.4	3.4	1.2	v	40-50	c	p	n	n
141-12	6.0	5.1	2.0	v	40-50	r	p	n	n
22-11	2.6	5.0	.9	v	60-70	c,i	s	n	y
138-13	5.3	3.0	1.3	v	60-70	c	p	n	n
141-14	3.6	3.1	.9	v	40-50	i	p	n	n
161-18	9.4	5.6	4.2	v	60-70	r	p	n	n
161-19	8.2	7.4	1.6	v	50-55	c	s	n	n
134-6	3.3	4.9	.8	v	-	-	-	n	y
141-20	4.3	5.3	.9	v	-	-	-	n	n

continued

Table 8. (continued)

TOOL NO.	L	W	T	B	SPINE ANGLE	EOS	EC	R	D
FISH BUTCHERING									
12-10	4.1	4.7	.5	v	15-20	v	p	n	y
12-14	2.4	3.0	.6	v	20-30	v	s	n	y
134-16	3.7	5.7	.8	v	40-45	s,v	s	n	n
20-7	4.1	3.9	.8	n	20-40	v,i	s	n	n
46-3	2.8	3.1	.9	v	40-50	i	s	n	n
138-12	5.8	3.3	1.8	v	15-25	v	s	n	y
134-17	5.0	5.4	.9	v	20-25	i	s	n	n
DEER BUTCHERING									
12-3	4.9	4.7	.7	v	10-25	v,i	p	n	n
12-6	6.2	5.2	.8	v	20-40	c	p	n	n
43-4	4.7	3.4	1.0	v	30-50	v	s	n	y
80-1	5.6	2.9	.6	v	25-40	v,i	p	n	-
80-12	6.0	4.0	1.4	v	40	s,v	s	n	-
80-13	3.4	3.6	.7	v	20-30	s	s	n	-
1-4	4.1	3.0	1.0	n	20-30	v	p	n	n
11-9	4.3	4.2	1.0	n	30	v	s	n	n
22-17	4.1	3.2	.6	v	20-25	c	s	n	n
27-5	3.2	2.2	.6	v	20-40	c	s	n	n
35-3	2.6	2.1	1.7	n	30-40	v	s	n	n
36-8	4.0	1.5	.5	v	25-30	i	p	n	y
45-7	3.3	4.1	.8	v	50	s	s	n	n
12-15	3.3	3.8	1.0	v	25-30	i	s	n	y
22-10	4.2	3.1	.7	v	30-40	s	s	n	y
22-15	6.5	3.8	1.0	v	20-30	i	p	n	n
45-6	3.6	4.9	1.0	v	20-40	i	s	n	y
12-11	8.8	5.4	1.5	v	50	v,c	p	n	y
80-3	7.4	6.2	2.6	v	40	v	s	n	-
80-14	3.7	4.7	.8	v	30	-	p	u	-
44-8	4.3	3.0	1.6	v	60-70	c	s	n	n
44-11	4.7	4.8	1.4	v	50-60	i	p	n	n
2-1	2.5	2.1	.4	n	20-40	s	s	n	n
2-3	2.1	2.5	.4	n	25-40	i	p	n	n
11-3	3.6	2.6	.7	n	30-40	i	p	n	n
12-1	2.8	3.0	.9	v	40-50	i	s	n	y

continued

Table 8. (continued)

TOOL NO.	L	W	T	B	SPINE ANGLE	EOS	EC	R	D
WOOD									
27-11	2.9	4.1	1.0	v	60-65	i	s	n	y
36-5	4.5	2.8	1.6	v	50-60	v	n	n	y
44-10	4.1	3.6	1.2	v	40-50	v	p	n	y
141-6	5.0	3.5	1.0	v	50	r	p	n	n
141-11	7.5	5.0	1.8	v	55-60	i	p	n	n
161-20	8.0	5.8	2.2	v	60	c	p	n	n
26-10	5.4	4.0	1.3	n	60	i	p	n	n
27-15	4.5	3.4	1.1	v	60-70	c	s	n	y
134-23	9.3	5.6	2.6	v	60-70	i	s	n	y
138-6	4.5	3.3	.9	v	40-60	v	p	n	n
138-16	2.3	5.8	.8	v	40-50	i	p	n	n
141-4	3.3	2.4	.7	v	50-60	c	s	n	n
141-17	5.9	4.4	1.5	v	60-70	r	p	n	n
27-10	2.9	3.5	.8	v	15-30	s	s	n	y
134-18	4.5	4.5	.7	v	20	s,v	s	n	n
63-2	3.5	3.6	.7	v	20-25	s	n	n	y
82-3	3.8	2.4	.7	v	20-30	v,c	p	n	-
27-7	3.3	2.2	.6	v	10-20	s	s	n	n
27-14	4.8	3.3	.7	v	10-20	v	s	n	y
12-5	6.3	4.8	.8	v	30-40	c	n	n	y
17-12	5.2	3.5	.7	v	20	r	s	n	n
80-9	3.4	2.5	.5	v	25-40	c	s	n	-
80-10	2.6	2.0	.4	v	10	c	s	n	-
80-11	2.5	2.4	.7	v	20-30		n	n	-
23-4	2.7	4.8	.6	n	20-25	i	n	n	n
27-8	3.6	2.3	.8	v	30-40	v	n	n	n
44-7	4.8	3.7	1.3	v	30-40	v	s	n	n
44-9	4.8	3.0	1.3	v	30	v	s	n	n
161-15	5.0	3.8	1.0	v	40-50	i	p	n	n

continued

Table 8. (continued)

KEY

L = length measured at a right angle from the striking platform to the distal end (in cm)

W = width, maximum distance at right angles to length (in cm)

T = maximum thickness (in cm)

B = type of basalt, either vitreous "v" or non-vitreous "n"

SPINE ANGLE = spine plane angle of used edge, range indicated in degrees

EOS = edge outline shape, either straight "s", concave "c", convex "v", recurved "r", or irregular "i"

EC = edge curvature, either not present "n", slight "s", or pronounced "p"

R = deliberate retouch, either not present "n", unifacial "u"

D = pre-use damage, either present "y" or not present "n"

Table 9. Summary of tool use experiments. The tools are ordered by contact material. A key to the abbreviations used is at the end of the table.

TOOL	EL	CM	PART	COND	PREP	B	ACT	ANGLE	PR	AP	DUR	STR
PLANTS												
12-9	f	r	ps	f,w,g	n	n	sl	40-50	h	l	5:47	200
12-12	f	r	ps	f,w,g	n	n	sl	70-90	h	l-m	22:38	1750
134-11	f	gs	ps	f,d	n	n	sl	90-120	h	m	13:28	1000
43-2	f	c	ps	y,d	n	n	sl	120-140	h	m	3:01	150
161-11	f	c	ps	y,d	n	n	sl	120-140	h	m	19:25	700
HIDE												
1-3	l	fd	hh	o,d,g	rd,u,d,r	c	sc	130-150	h	m-h	5:15	500
12-13	l	fd	hh	o,d,g	rd,u,d,r	c	sc	80-110	hf	m-h	38:05	4000
17-14	l	fd	hh	o,d,g	rd,u,d,r	c	sc	70-120	hw	m-h	47:04	5200
22-7	l	fd	hh	o,d,g	rd,u,d,r	c	sc	55-65	h	m-h	1:32	200
23-3	l	fd	hh	o,d,g	rd,u,d,r	c	sc	80-120	h	m-h	10:48	1600
27-16	l	fd	hh	o,d,g	rd,u,d,r	c	sc	135-150	h	l-m	18:05	2000
39-1	l	fd	hh	o,d,g	rd,u,d,r	c	sc	80-100	hf	m	30:39	2350
44-4	l	fd	hh	o,d,g	rd,u,d,r	c	sc	70-85	h	m-h	8:00	1000
3-3	l	fd	fh	f	rd,u,r	c	sc	90-115	hf	m-h	14:26	1000
3-4	l	fd	fh	f	rd,u,r	c	sc	85-105	h	l-h	7:22	1000
22-12	l	fd	fh	f,g	rd,u,r	c	sc	90-110	hf	m-h	16:23	1000
27-17	l	fd	fh	f	rd,u,r	c	sc	90-120	hf	m-h	58:04	2537
31-2	l	fd	fh	f	rd,u,r	c	sc	90-120	hf	m-h	4:45	250
44-6	l	fd	fh	f	rd,u,r	c	sc	75-90	hw	m-h	40:00	1750
46-4	l	fd	fh	f	rd,u,r	c	sc	100-120	hf	m-h	38:45	3200
80-4	g	md	fh	f	rd,n	f	sc	75-90	h	m-h	5:00	-
80-5	g	md	fh	f	rd,n	f	sc	75-90	h	m-h	5:00	-
80-6	g	md	fh	f	rd,n	f	sc	-	h	-	:30	-
80-7	g	md	fh	f	rd,n	f	sl	-	h	-	:30	-
80-8	g	md	fh	f	rd,n	f	sl	-	h	-	1:00	-
12-17	l	fd	h	o	rd,u,r,d,h	n	sa	45-90	h	m-h	4:16	400
134-9	l	fd	h	o	rd,u,r,d,h	n	sa	40-60	h	m-h	18:02	1200
BONE												
21-6	l	fd	b	f,d	d,f	n	sa	85-95	h	h	1:40	200
26-8	l	fd	b	f,d	d,f	n	sa	85-95	h	h	1:34	400
26-9	l	fd	b	f,d	d,f	n	sa	80-100	h	h	2:38	400
46-2	l	fd	b	f,d	d,f	n	sa	80-100	h	h	4:05	400
11-6	l	fd	b	f,w	d,s	n	sa	80-90	h	h	1:09	200
18-19	l	fd	b	f,w	d,s	n	sa	80-90	h	h	:55	200
22-20	l	fd	b	f,d	d,f	n	sc	120-140	h	h	2:18	70
17-13	l	fd	b	f,w	d,s	n	sc	70-90	h	m-h	1:39	100
63-1	l	fd	b	f,d	d,f	n	w	30-50	h	h	1:04	60
138-19	l	fd	b	o,d	n	n	g	0-45	h	h	11:41	600

continued

Table 9. (continued)

TOOL	EL	CM	PART	COND	PREP	B	ACT	ANGLE	PR	AP	DUR	STR
ANTLER												
43-3	l	fd	a	f,d,g	n	n	sa	85-95	h	h	:38	200
134-22	l	fd	a	f,d,g	n	n	sa	70-80	h	h	1:40	400
141-7	l	md	a	o,d,g	n	n	sa	85-95	h	h	:45	200
11-2	l	fd	a	f,w	s	n	sa	85-95	h	m-h	1:22	200
11-12	l	md	a	o,w	s	n	sa	85-90	h	h	1:10	200
17-11	l	fd	a	f,w	s	n	sa	85-95	h	h	2:00	400
27-3	l	fd	a	f,w	s	n	sa	85-95	h	h	1:00	200
36-7	l	fd	a	f,w	s	n	sa	85-95	h	h	1:10	200
18-17	l	md	a	o,w	s	n	sa	85-95	h	h	1:10	200
141-12	l	md	a	o,w	s	n	sa	85-95	h	h	2:34	400
22-11	l	md	a	o,d,g	n	n	sc	40-60	h	h	1:03	100
138-13	l	fd	a	f,d,g	n	n	sc	40-60	h	h	:56	115
141-14	l	fd	a	f,w	s	n	sc	60-80	h	h	1:14	100
161-18	l	md	a	o,w	s	n	sc	60-75	h	h	2:14	200
161-19	l	fd	a	f,w	s	n	sc	60-80	h	h	2:13	200
134-6	l	fd	a	o,w	s	s	b	80-100	h	m-h	15:10	1000
141-20	l	fd	a	o,w	s	s	g	10-50	h	m-h	20:45	1000
FISH												
12-10	l	t	f,b,s,k	f,w	f,s,g	l	sl	50-70	h	l-m	5:20	500
12-14	l	t	f,b,s,k	f,w	f,s,g	l	sl	80-100	h	m-h	4:43	200
134-16	l	t	f,b,s,k	f,w	f,s,g	l	sl	70-110	h	m-h	9:35	519
20-7	l	t	f,b,s,bk	f,w	f,s,g,p	l	sl	80-100	h	l-m	3:52	500
46-3	l	t	f,b,s,bk	f,w	f,s,g,p	l	sl	85-95	h	l-m	2:06	200
138-12	l	t	f,b,s,bk	f,w	f,s,g,p	l	sl	80-100	h	l-m	9:43	1000
134-17	l	t	k,s	f,w	f,s,g	l	sl	40-50	h	l-m	4:16	500
DEER												
12-3	b	fd	f,hi,b	f	g	n	sl,sa	75-95	h	l-m	28:23	750
12-6	b	fd	f,hi,b	f	g	n	sl,sa	10-100	h	m-h	38:40	2820
43-4	b	fd	f,hi,b	f,g	g	n	sl,sa	30-100	h	h	13:30	925
80-1	g	md	f,hi	f	g	n	sl	-	h	m-h	20:00	-
80-12	g	md	f,hi	f	g	n	sl,sa	-	h	-	50:00	-
80-13	g	md	f,hi	f	g	n	sl,sa	-	h	-	5:00	-
1-4	b	fd	f	f	g	n	sl	50-70	h	l-m	9:27	700
11-9	b	fd	f	f	g	n	sl	50-90	h	l-m	9:23	466
22-17	b	fd	f	f	g	n	sl	80-90	h	l-m	1:44	100
27-5	b	fd	f	f	g	n	sl	75-90	h	l-m	2:05	150
35-3	b	fd	f	f	g	n	sl	60-90	h	l-m	2:50	165
36-8	b	fd	f	f	g	n	sl	80-100	h	l-m	1:08	50

continued

Table 9. (continued)

TOOL	EL	CM	PART	COND	PREP	B	ACT	ANGLE	PR	AP	DUR	STR
DEER (continued)												
45-7	b	fd	f,b	f	g	n	sl,sa	80-100	h	m	34:22	2020
12-15	b	fd	f,t,b	f	g	n	sl	40-70	h	l-m	6:37	150
22-10	b	fd	f,hi,c,b	f	g	n	sl	10-100	h	m-h	12:09	450
22-15	b	fd	f,hi,c,b	f	g	n	sl	75-100	h	m-h	15:22	350
45-6	b	fd	f,t,b	f	g	n	sl	60-100	h	m-h	20:15	740
12-11	b	fd	f,hi,c,b	f,g	g	n	sa,sl	65-105	h	h	32:04	920
80-3	g	md	f,hi,t,c,b	f	g	n	sa	70-100	h	m-h	3:00	-
80-14	g	md	f,hi,t,c,b	f	g	n	sa	-	h	h	5:00	-
44-8	l	fd	f,b,p	f,d	n	n	sc	80-100	h	m-h	28:24	1000
44-11	l	fd	p,t,f,b	f,d	n	n	sc	80-100	h	m-h	6:45	300
2-1	l	fd	t,c,p,b	f,d	n	n	sl	60-100	h	m-h	9:53	350
2-3	l	fd	f,b,p	f,d	n	n	sl	35-60	h	m-h	9:46	350
11-3	l	fd	f,b	f,d	n	n	sl	35-90	h	m-h	5:00	210
12-1	l	fd	f,b,p	f,d	n	n	sl	10-45	h	m-h	17:40	1000
WOOD												
27-11	l	d	ba,w	o,d	b	n	sc	90-100	h	m	9:44	1150
36-5	l	bp	ba,w	o,d,g	b	n	sc	120-130	h	m-h	17:39	500
44-10	l	cc	ba,w	o,d	b	n	sc	80-100	h	m	4:38	500
141-6	l	s	ba,w	o,d	b	n	sc	85-95	h	m-h	5:03	500
141-11	l	cc	ba,w	o,d	b	n	sc	60-90	h	m-h	12:34	1000
161-20	l	s	w	o,d	b,ba	n	sc	80-90	h	m-h	5:13	500
26-10	l	s	ba,w	f,d	b	n	sc	80-90	h	m-h	7:58	500
27-15	l	s	ba,w	f,d	b	n	sc	80-90	h	m-h	18:00	2050
134-23	l	s	ba,w	f,d	b	n	sc	75-105	h	m-h	68:58	5850
138-6	l	d	ba,w	f,d	b	n	sc	85-95	h	m-h	4:47	500
138-16	l	s	ba,w	f,d	b	n	sc	85-95	h	m-h	4:51	500
141-4	l	s	ba,w	f,d	b	n	sc	95-105	h	m-h	1:46	200
141-17	l	s	w,ba	f,d	b	n	sc	85-95	h	m-h	6:19	500
27-10	l	s	w	f,d	b,ba	n	sa	85-95	h	h	1:55	400
134-18	l	s	w	f,d	b,ba	n	sa	85-95	h	h	6:30	1000
63-2	l	s	w	o,d	b,ba	n	sa	80-100	h	h	2:51	400
82-3	a	p	w	o,d	b,ba	n	sa	-	h	-	4:00	-
27-7	l	b	ba	o,d	rt	n	sa	80-100	h	m	6:00	600
27-14	f	b	ba	o,d	n	n	sl	80-100	h	m-h	6:09	150
12-5	l	s	ba,w	f,d	b	n	w	25-35	h	m-h	8:10	500
17-12	l	s	ba,w	f,d	b	n	w	40-55	h	m-h	2:59	200
80-9	f	w	ba,w	f,d	b	n	w	-	h	-	7:00	-
80-10	f	w	ba,w	f,d	b	n	w	-	h	-	2:00	-
80-11	f	w	ba,w	f,d	b	n	w	-	h	-	12:00	-
23-4	l	s	w,b	f,d	b	n	w	120-140	hw	m-h	2:38	200

continued

Table 9. (continued)

TOOL	EL	CM	PART	COND	PREP	B	ACT	ANGLE	PR	AP	DUR	STR
WOOD (continued)												
27-8	l	bp	ba,w	o,d,g	b	n	w	20-40	h	m	13:22	550
44-7	l	bp	w	o,d	b,ba	n	w	20-40	h	m	13:06	420
44-9	l	bp	w	o,d	b,ba	n	w	20-40	h	m	4:02	200
161-15	l	s	w	o,d	b,ba	n	w	45-55	h	h	5:25	500

Key

EL = location of experiment: field "f", lab "l", barn "b", garage "g", apartment "a"

CM = contact material: reeds "r", wild grass and sedge "gs", cattails "c", fallow deer "fd", mule deer "md", trout "t", dogwood "d", balsam poplar "bp", choke cherry "cc", Saskatoon "s", white birch "b", willow "w", ponderosa pine "p"

PART = part of contact material worked: plant stems "ps", hair side of hide "hh", flesh side of hide "fh", hide lacking hair or flesh "h", intact hide "hi", antler "a", bone "b", periosteum "p", cartilage "c", flesh and fat "f", tendons "t", fish scales "k", fish skin "s", wood "w", bark "ba", backing "bk"

COND = contact material condition: fresh "f", old "o", wet "w", dry "d", gritty "g"

PREP = modification or preparation of contact material prior to experiment: soaked in water "s", soaked in urine "u", rinsed with water "r", defleshed "d", dehaired "h", removed from deer "rd", frozen and thawed "f", gutted "g", ribs and spine removed "p", branches removed "b", bark removed "ba", removed from tree "rt", no preparation "n"

B = backing: aspen logs "l", cardboard over a plastic sheet "c", cement floor "f", soft leather sheet "s", no backing "n"

ACT = tool action: slicing "sl", sawing "sa", scraping "sc", whittling "w", graving "g", boring "b"

ANGLE = angle between the ventral edge aspect of a tool and the contact material surface in degrees (range)

PR = prehension of tool: hand-held "h", wrapped in rawhide and hand-held "hw", hafted onto an antler handle "hf"

AP = pressure applied to tool: light "l", light to medium "l-m", medium "m", medium to heavy "m-h", heavy "h"

DUR = duration of experiment in minutes and seconds (min:sec)

STR = number of tool strokes

character. The tasks or "goals" included skinning a deer, dehairing a hide, filleting fish and preparing arrow shafts, among others. However, tasks were not necessarily carried out to the stated goal with every tool, and in many cases several tools were used to accomplish a single goal. For example, several tools were used to butcher a deer, while it may have been possible to do this with one or two tools. Sometimes tools were used in a goal-oriented fashion without achieving the ultimate goal; for example, parallel grooves were engraved in an antler as in the "groove and splinter" technique, but the experiment did not extend to prying out a splinter (see also Vaughan 1985:15).

Most of the experiments were undertaken in 1984 and 1985, although a few from 1980 and 1982 were used in this study. The early experiments did not involve the recording of as many variables as the later ones and these differences are noted where appropriate below. Tools used in 1980 and 1982 are identified by the prefix "80" or "82".

5.2. *Pre-Experiment Observations*

5.2.1 *Procedure*

Tools were prepared for use in experiments as follows: (1) a flake was manufactured and labelled, (2) it was carefully cleaned, (3) it was examined microscopically for accidental damage, (4) pertinent attributes of the flake were recorded, (5) it was stored for use in a future experiment. Contact materials were either collected and stored for later use in an experiment, or they were encountered and used in experiments in the field. In either case pertinent attributes of a contact material were recorded prior to its use in an experiment.

5.2.2 *Recording Methods and Equipment*

Pertinent formal attributes of tools, their measurements and illustrations, as well as pre-use microscopic observations and photo numbers were all recorded on one standardized form. Examination of the edges for accidental damage was undertaken with either a Wild Heerbrugg zoom stereomicroscope (12-100 x) or a Leitz Wetzlar stereomicroscope (10-100 x). Photomicrographs of accidental damage were taken on a Zeiss stereomicroscope (70 x) equipped with a built-in 35 mm camera using Kodak Plus-X black and white film (ASA 400).

Pertinent characteristics of a contact material were recorded immediately prior to experimentation on the same form used for recording observations during an experiment.

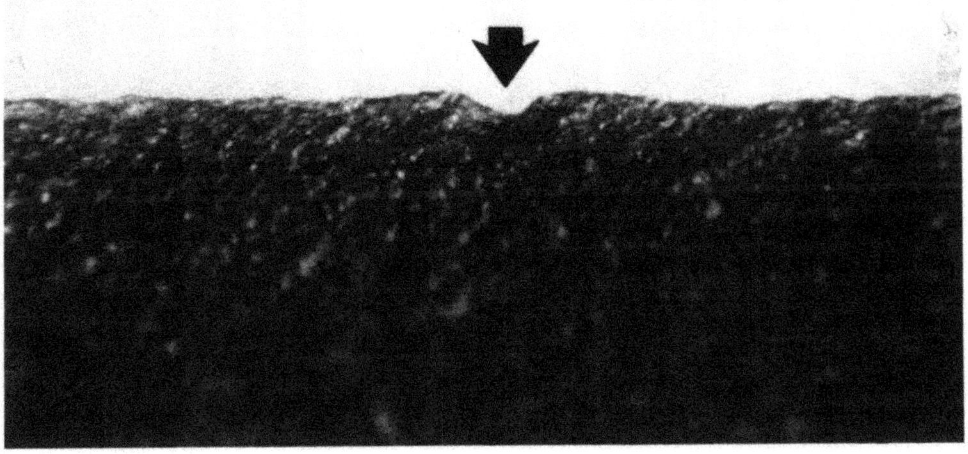

Figure 16. Slight pre-use damage on the ventral edge-aspect of tool 39-1. The arrow points to an isolated feather-terminated microflake scar. The tool was later used in a hide dehairing experiment.
(245 x)

Figure 17. Severe pre-use damage on the ventral edge-aspect of tool 43-11. Several overlapping and superposed step-terminated microflake scars are visible. Damage was severe enough to prevent this tool from being used in an experiment.
(245 x)

5.2.3 *Manufacture of Tools*

Flakes were produced by direct freehand percussion of basalt nodules with a hammerstone, or occasionally, an elk antler billet. In a few cases bipolar percussion involving a hammerstone and anvil was employed. Flakes potentially usable as tools (i.e. not blocky, reasonably large, and sharp-edged) were handled carefully and set aside. However, during detachment from the core some flakes underwent spontaneous retouch (Newcomer 1976), hit other flakes or forcefully hit the cardboard backing and thereby sustained minor edge damage. Flakes intended for use as tools in experiments were labelled by core-nodule number and an arbitrary sequential number (e.g. flake "163-11" = core 163, flake 11) and set aside on cardboard trays.

5.2.4 *Accidental Damage*

Usable edges were examined with a stereoscopic microscope for pre-use damage at magnifications up to 100 x (Figures 16,17). Those with more than slight edge-damage (microflake scars) were rejected as potential tools (Figure 17). Minor damage was described, its location indicated on a sketch of the flake, and in many cases photographed (Figure 16). After examination, potential tools were wrapped in soft paper tissue and stored in labelled paper bags.

5.2.5 *Tool Attributes*

A number of basic attributes were recorded for each potential tool, with special attention given to describing usable edges: quality of basalt, length, width and thickness of the tool, spine-plane angle (Tringham et al. 1974:179), edge outline shape (Keeley 1980:19-20), amount of edge curvature (Tringham et al. 1974:3), and the presence of deliberate retouch (Table 8).

5.2.6 *Contact Material*

Contact material, also known as "worked material" (Tringham et al. 1974; Keeley 1980) or "object material" (Odell and Odell-Vereecken 1980), is the matter worked with the tool. In this study the following contact material classes were worked: bone, antler, hide, flesh, wood, and plants. Various parts of a contact material have qualities which could differentially condition the development of microwear. Therefore, the specific part of a contact material worked with a tool was recorded; for example, plant *stems* were cut rather than heads, leaves or roots, and the *flesh* or *hair* sides of hides were worked in different experiments (Table 10). Some contact materials were used in their natural state, while others were modified. In some cases this was done deliberately to mirror aboriginal behaviour, such as soaking bone or antler in water to soften it (Teit 1909:474-475). In other cases it was done out of necessity, such as freezing fresh material for later experimentation, although it is likely that material which was frozen and then thawed was also used prehistorically.

Most observations regarding contact materials were recorded immediately prior to the beginning of an experiment. Occasionally, however, some materials were contacted unexpectedly in the course of an experiment. Examples would be the accidental contacting of backing while scraping hide, or the incidental contacting of bone or cartilage in a flesh cutting experiment. The frequency of contacting these materials during an experiment was recorded.

5.3 *Observations During Experiments*

5.3.1 *Procedure*

Experiments were conducted as follows: (1) a flake with an edge judged suitable for the proposed task was selected from those available; (2) the contact material to be worked was made ready; (3) the experiment began and was recorded on a prepared form and with photographs; (4) the experiment ended and further observations were recorded; and (5) the tool was carefully stored for future examination.

5.3.2 *Recording Methods and Equipment*

All aspects of the experiments were recorded on prepared forms before, during and immediately after a tool-use episode. The used part of a tool was indicated on a sketch on the form, as were the details of tool prehension and its positioning and motion relative to the contact material during use. While the experiments were in progress the duration of tool-use was recorded with a digital stop watch. Photographs were taken during experiments in progress with a Nikon FE 35 mm camera using Kodak Plus-X black and white film (ASA 125) or Kodachrome colour slide film (ASA 100) (Figures 18-23).

5.3.3 *Experimental Conditions*

The experiments were conducted in both lab and field situations. All experiments involving green plants, and some involving wood working, were conducted in completely natural environments. Several other experiments were conducted on the cement floors of a barn and a garage, representing neither completely natural nor laboratory conditions (Figures 18-23). By taking the experiments indoors, an element of authenticity was lost, particularly the degree of "grittiness" of the work area (Keeley 1980:15; Tringham et al. 1974:184). The lab was considerably less gritty than outdoors, but the barn and garage floors

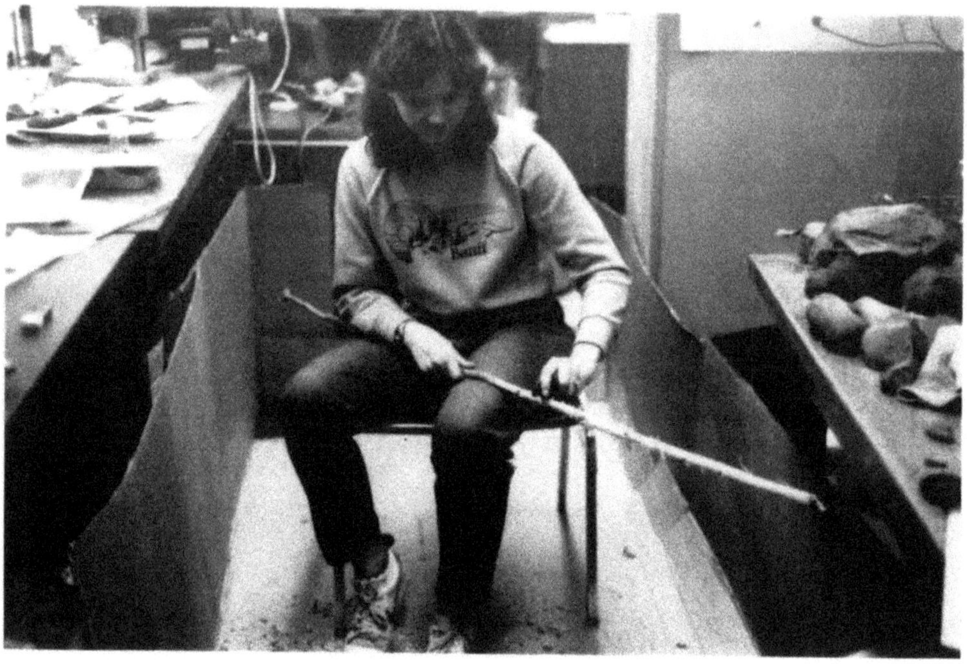

Figure 18. Tool 134-23 being used to scrape a Saskatoon branch by Joanne Hadar.

Figure 19. Close-up view of tool 134-23 during use.

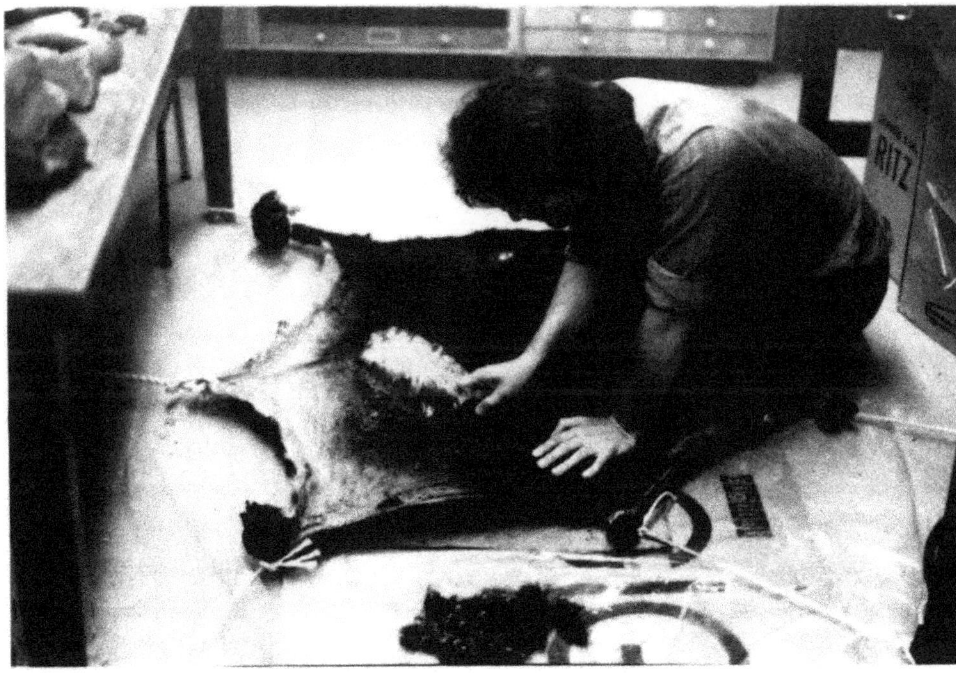

Figure 20. Author using tool 27-16 to scrape hair from a fallow deer hide.

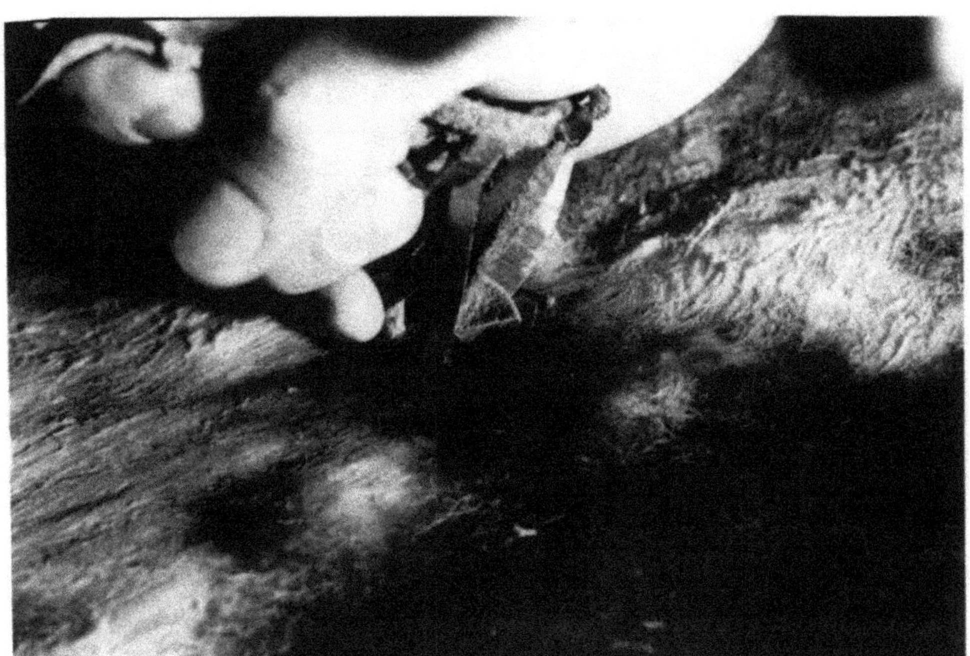

Figure 21. Close-up view of tool 46-4 during use as a hafted scraper during the final stages of defleshing a fallow deer hide.

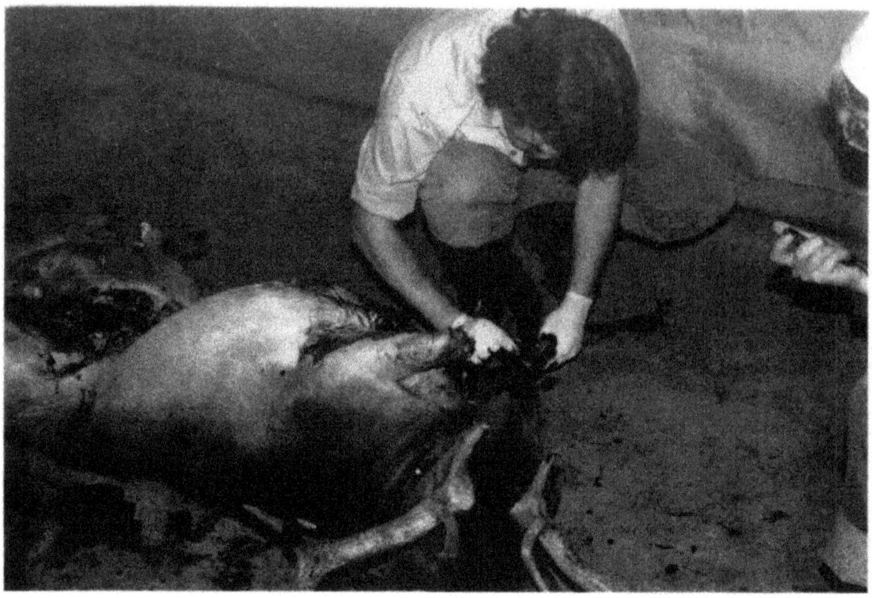

Figure 22. Author using tool 12-11 to cut off a fallow deer forelimb. At the extreme right Kit Krozser can be seen timing the experiment and counting strokes.

Figure 23. Kit Krozser using tool 1-4 to slice through fallow deer fat and flesh.

were only slightly less gritty than many outdoor situations. Airborne particles which may come between the tool and contact materials were largely excluded in the indoor experiments. However, materials used in the experiments were not deliberately cleaned to remove grit, and substantial amounts of grit (sand and silt) were present on several contact materials, most notably on the hair side of the fallow deer hide and on fallow deer antlers (adhered by dried blood).

5.3.4 Experimenters

All experiments and observations undertaken in 1980 and 1982 were carried out and recorded solely by the author, as were most of those undertaken in 1984 and early 1985. However, two individuals (Kit Krozser, Joanne Hader) participated in eleven experiments under the author's direction in 1984 and 1985, and several people (Kit Krozser, Joanne Hader, Sharon Mueller, Leslie Amundson) assisted in the recording of observations, especially elapsed time and number of strokes, and by taking photographs.

5.3.5 Backing

Backing refers to material on which the contact material is directly rested while being worked. Materials worked while being hand-held or partly resting on the experimenters leg are considered not to have been backed (Figures 18,19). The backing was rarely contacted with stone tools during the experiments, but this was recorded when it occurred. Aspen logs, a soft leather sheet, cardboard and plastic were the backings used in various experiments (Figure 20). Only the logs were contacted to any appreciable degree during the experiments.

5.3.6 Prehension

Only a few tools used in hide-scraping experiments were hafted onto a mule deer antler handle (Table 9). Tools were partially wrapped with thin, soft leather to prevent slippage, and were bound to the haft with thin leather strips (Figure 21). Some hand-held tools were partially wrapped in rawhide to spare the hand. It was necessary to deliberately "back" most tools by chipping with a hammerstone and to dull sharp edges or projections by grinding before they could be comfortably used with a bare hand.

5.3.7 Tool Action

Tool "action" refers to the "activity" (Keeley 1980:17-19) or "direction" (Tringham et al. 1974:181) of the tool during use. Six tool actions were

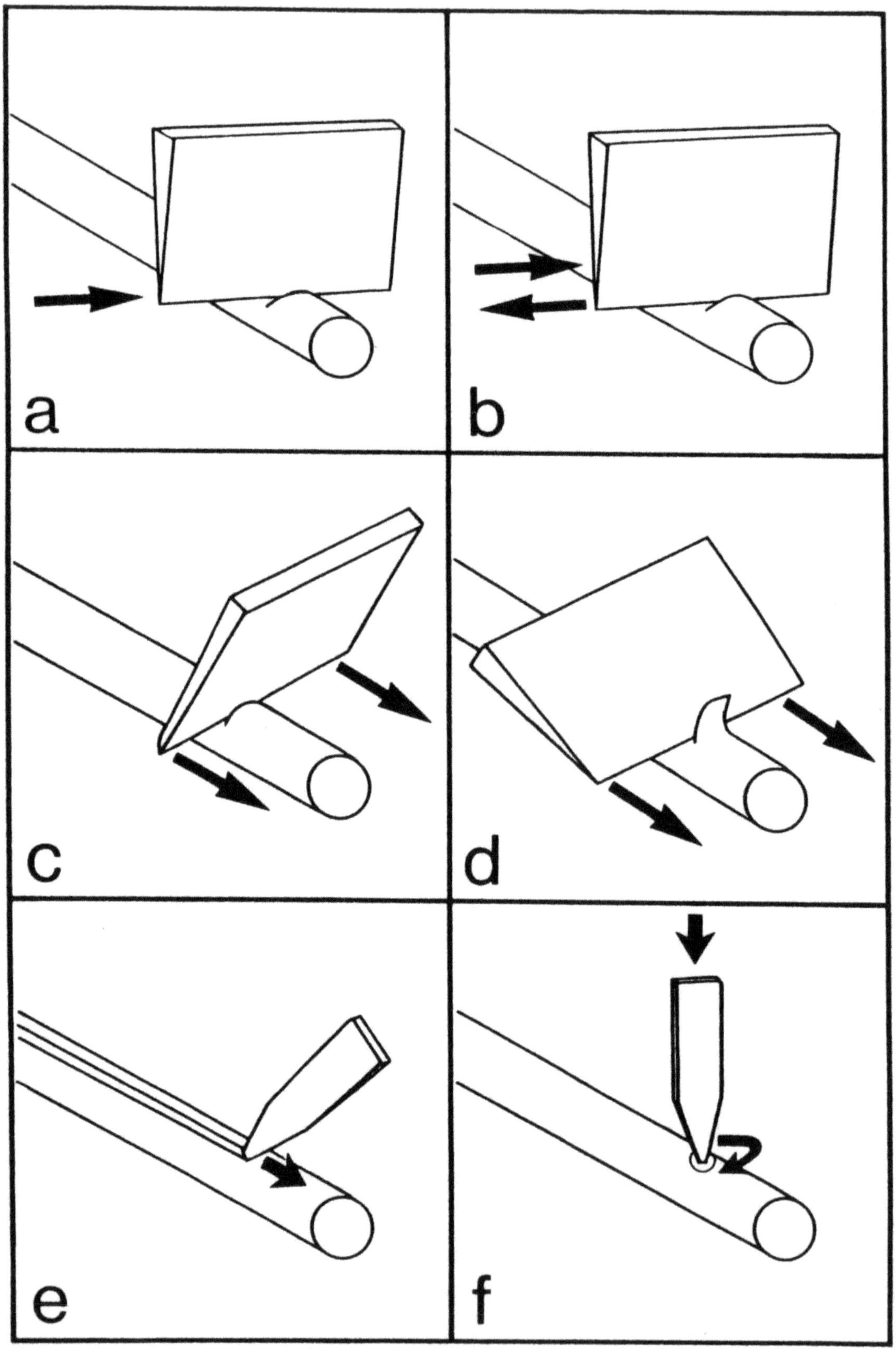

Figure 24. Tool actions: a-slicing, b-sawing, c-scraping, d-whittling, e-graving, f-boring.

employed in the experiments: slicing, sawing, scraping, whittling, graving and boring (Figure 24). Tringham, Odell (Tringham et al. 1974:181; Odell and Odell-Vereecken 1980:98-100) and others make a useful distinction between motions longitudinal to the tool-edge and those transverse to the edge.

For the present experiments the following definitions of tool actions are used. *Slicing* ("cutting" in the Tringham-Odell terminology, "slicing" in Keeley's terminology) refers to a one-way longitudinal motion undertaken with the tool edge at a high angle (in the vicinity of 90 degrees) to the contact material (Figure 24a). *Sawing* ("sawing" or "cutting" according to Keeley; "sawing" according to Tringham and Odell) refers to a back and forth longitudinal motion undertaken with the tool edge also usually at a high angle to the contact material (Figure 24b). *Scraping* is a one-way movement with the tool edge being drawn transversely and usually at high angles to the contact surface (Figure 24c). Movement of the tool can be either towards or away from the user, as long as the same edge leads in either direction. *Whittling* refers to a one-way movement of a tool, also moved transversely across the contact material, but at a very acute angle and always away from the tool user (Figure 24d). As pointed out by Odell (Odell and Odell-Vereecken 1980:99), *graving* can be either a longitudinal or transverse action. The longitudinal/ transverse classification is really only useful for describing tool use involving *edges,* and is not really appropriate for actions involving points or spurs. *Graving* is a one-way movement of a spurred or pointed tool, with the point drawn or pushed across the contact surface (Figure 24e). *Boring* is simultaneously a semi-circular, bi-directional and downward movement of a pointed or spurred tool held at or near 90 degrees to the contact surface (Figure 24f).

5.3.8 Applied Pressure

The pressure or force applied to tools used for different activities was variable (see Table 9) and often the amount of pressure applied to a tool changed during the course of a single experiment. This was mainly caused by the increasing pressure needed to get results from a dull tool. No attempt was made to keep the pressure artificially constant (as with Tringham et al. 1974:182) and it varied as was necessary to accomplish a given task. Applied pressure was generally not recorded during the 1980 and 1982 experiments.

5.3.9 Angle of Tool to Contact Surface

The angle between the ventral edge-aspect of an experimental tool and the contact surface was measured several times during each experiment ("angle of the edge to the worked material" in the Tringham-Odell terminology and "angle of attack" according to Keeley). An optimum angle was quickly established with each tool, although this sometimes changed during the course of an experiment as the edge became duller or the edge angle changed due to attrition, or if the shape of the contact material was altered. No effort was made to keep this angle artificially constant throughout an experiment (as with Tringham et al. 1974:181-182), as the concern was with using the tool in an effective and realistic manner. Even so, the tool angle usually only varied 10 to 20 degrees, and often mainly within a 5 to 10 degree range. This angle was usually not recorded during the 1980 and 1982 experiments.

5.3.10 Duration of Tool Use

Some researchers use only "elapsed time" (Keeley 1980:8) or "number of strokes" (Tringham et al. 1974:184) to measure the duration of an experiment. Both are used in this study to insure the widest possible comparability with other research. Time was measured by stopwatch and was stopped during the experiment while various observations were recorded, or for a break during longer experiments. Only elapsed time was recorded for the 1980 and 1982 experiments.

5.3.11 Post-Experiment Tool Treatment

At the end of an experiment the used tool was wrapped in tissue or soft paper towel and stored in a labelled paper bag. The bagged tool was protected from possible post-use damage by being stored in a cabinet and only taken out when necessary for examination. Eventually, all of the tools were stored in small, clear plastic bags for ease of access.

CHAPTER 6

DEPENDENT VARIABLES, RECORDING METHODS, AND PROCEDURES

6.1 *Introduction*

The dependent variables in this series of experiments are wear traces only clearly observable with the aid of light microscopes or a scanning electron microscope (SEM). The types and general characteristics of these variables are described below, as are the methods and procedures used to record them.

6.2 *Analytical Procedure*

A tool used in an experiment was removed from storage and prepared for analysis with a light microscope in the following manner: (1) each tool was gently brushed with a very soft tooth brush in warm, soapy water to remove residues from the contact material; (2) after air drying, each tool was placed in a clean plastic bag (baby bottle liner); (3) immediately prior to examination under the microscope the tool edge was gently wiped with alcohol or acetone to remove finger grease (this was repeated when necessary during the examination); (4) the tool was examined with a microscope, and microwear observations were recorded and photographed; and (5) after examination the tool was placed in its bag and stored until it was needed again.

The procedure for tools to be examined with a SEM was as follows: steps (1),(2) as above; (3) a segment of the used edge of the tool was removed with a saw; (4) both the tool and segment were cleaned with soapy water; (5) after drying, the tool was replaced in its bag and stored; (6) the segment was placed in an ultrasonic cleaner; (7) the segment was allowed to dry, placed in a sterile plastic bag and transported to the SEM; (8) the segment was adhered to a "stage" and coated with a thin layer of gold; (9) the segment was examined with the SEM and observations were made on the microwear and photographs were taken; and (10) the tool segment was placed in a bag and stored.

6.3 *Recording Methods*

All dependent variables were recorded in detail on forms prepared for this purpose. Each form was designed so that many observations could be recorded by merely checking off the appropriate box; however, there were no limits on recording observations which did not fit one of the predetermined categories. The microscope and magnifications used for the observations were recorded on the forms. Photomicrograph numbers were also recorded on the forms and in a photo catalog.

6.4 *Equipment*

A variety of microscopes were used to observe the dependent variables. To study microflake scars and edge rounding, a Zeiss Jena stereomicroscope with a range of magnifications between 16 x and 100 x was used. However, the microflake scar characteristics were recorded as they appeared at 40 x to standardize observations. Two microscopes were used for the observation of micropolishes and striations. The first, a Nikon Optiphot metallurgical microscope equipped for incident light observation and photography had a range of magnifications between 50 x and 400 x. It was fitted with a video-camera and magnifications of up to 1000 x could be obtained on a monitor, although the images were quite fuzzy. This microscope proved unsuitable for observing microwear polishes and only a few tools were examined (these were later re-examined with the Unitron). A much more suitable microscope was a Unitron "N" metallurgical microscope with incident light, a xenon light attachment and a range of possible magnifications between 50 x and 600 x. It was equipped with a 35 mm camera, as well as a polaroid camera.

A Phillips 505 scanning electron microscope was used to examine parts of 23 tools and 3 unused flakes. Black and white photomicrographs were taken with Polaroid Type 665 Positive/Negative Land Pack Film.

6.5 *Dependent Variables*

6.5.1 *Introduction*

Several different classes of wear, including microflake scars, polish, striations and rounding, have been commonly observed on used tools by researchers involved with the study of microwear (e.g. Semenov 1964, Tringham et al. 1974, Keeley 1980, and many others). All of these classes of wear have been observed on tools used in the present series of experiments. The characteristics of each class of wear are described below.

6.5.2 *Microflake Scars*

Microflake scars are found on the edge-aspects of a tool, not on the edge. Edge-aspects are the areas of the dorsal and ventral faces immediately adjacent to the edge (Figure 25). Snap-terminated microflake scars represent a part of the edge which has snapped off, however, these scars are usually oriented to one or the other edge-aspect to some degree.

Following previous research (Tringham et al. 1974; Cotterell et al. 1979) microflake scars are classed as having snap-, step-, hinge- or feather-terminations. Each microflake scar was recorded as being located on the dorsal or ventral edge-aspect (Figure 25). The ratio of scars on the dorsal vs ventral aspect was calculated (number of ventral scars divided by the number of dorsal scars = ventral:dorsal ratio) as it was considered a possibly useful index of the relative distribution of scars and potentially valuable for distinguishing tools on the basis of their action.

The spatial relationships between microflake scars or "configuration" was also recorded for each edge-aspect. When few non-overlapping, widely spaced microflakes scars occur on an edge-aspect, the configuration is described as "scattered" (Figure 26a). Continuous microflake scars, either not overlapping each other or only slightly overlapping, are described as a "contiguous" distribution (Figure 26b). Continuous, overlapping microflake scars comprise an "overlapping" configuration (Figure 26c). The frequent occurrence of small scars within larger scars is described as a "superposed" configuration (Figure 26d). Several of these configurations can be present on the same used edge-aspect.

Length of microflake scars was used as the measurement of size, although some previous researchers have used scar *width* to categorize scars by size (e.g. Keeley 1980:24-25). The major drawback in using width is that scars often overlap each other, thus truncating part of the width. Measurements of width are therefore not likely to be comparable between scars occurring in overlapping clusters and those occurring singly. In the present study, microflake scars were divided into four arbitrary size classes suggested by subjective evaluation of scar size distributions on several used tools. Obtaining precise measurements for

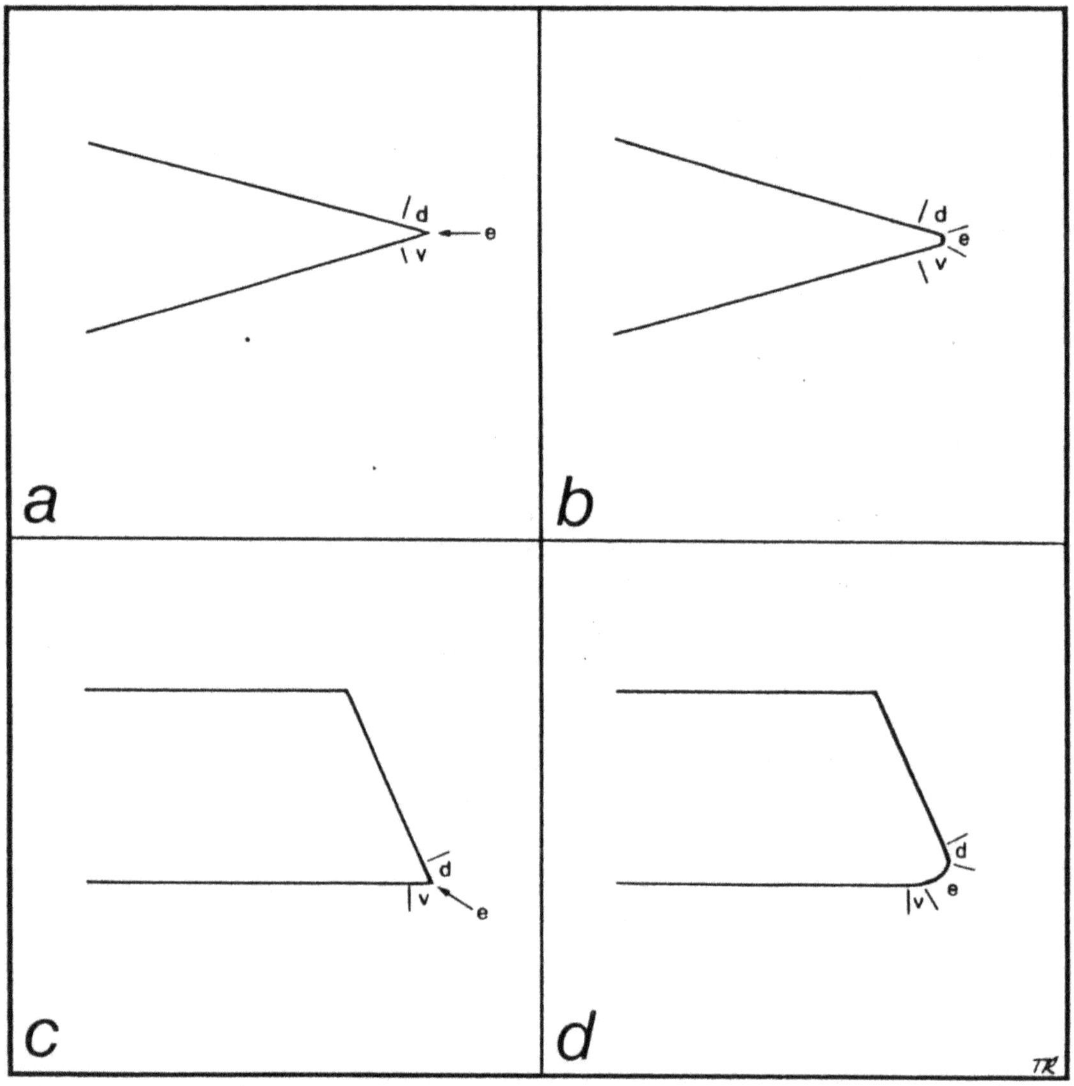

Figure 25. Relationship of the dorsal (d) and ventral (v) edge-aspects to the edge (e) of a tool. a. unrounded, low-angled edge, b. rounded, low-angled edge, c. unrounded, steep-angled edge, d. rounded, steep-angled edge.

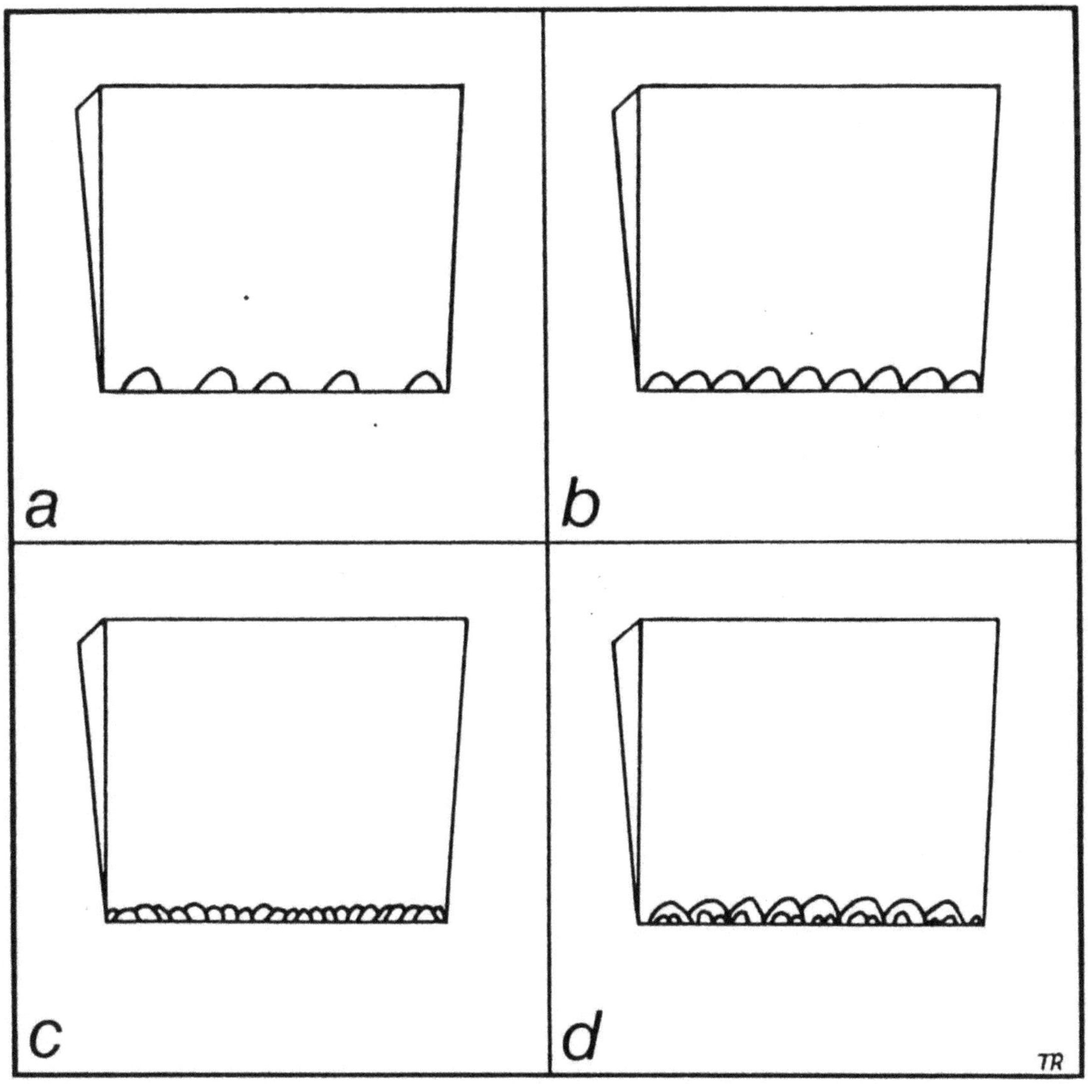

Figure 26. Spatial relationships between microflake scars.
a. scattered, b. contiguous, c. overlapping, d. superposed.

each flake scar would be extremely time consuming, considering the number of scars on the tools--they ranged from a few to over 100 scars and most had around 50. In this study small scars measure less than 1 mm in maximum length, medium scars between 1 and 2 mm, large between 2 and 3 mm, and very large scars are greater than 3 mm. Snap-terminated scars are recorded according to their length parallel to the edge (i.e. width) because this is the most realistic measure of size for these scars. With the exception of snap scars, very large scars rarely occur as a result of use.

Microflake scars were recorded as they appeared at 40 x magnification with a stereomicroscope, although higher magnifications were occasionally used. This baseline was set to insure comparability in observations between tools.

6.5.3 *Rounding*

Edges, microflake scar ridges, and prominent parts of the microtopography can become rounded by use. Rounded parts of a tool can be accompanied by abrasion, smoothing or polishing of the rounded surface. At very high magnifications, most used edges appear rounded to some degree, and some "rounded" edges are actually facetted. Descriptions of edge-rounding in Chapter 7 refer to observations made during the low magnification study of microflake scars. Edge-rounding was subjectively recorded as being slight, moderate or pronounced. Rounding of other aspects of a used tool was also recorded.

Edge-rounding on transversely used tools is commonly described as being oriented to the "leading" or "trailing" edge-aspects (Figure 27). This simply refers to the aspects which were "leading" or "trailing" during use. With the scraping actions used in the present study, the leading aspect was always the ventral aspect due to the shape of the flakes which were used.

6.5.4 *Polish*

Following others who have distinguished many diagnostic features of use-produced micro-polishes (especially Keeley 1980), great care was taken to record the polishes observed on the basalt tools. Some "polishes" were indeed smooth and shiny as one expects a polish to be. Other modifications of the tool surface resulted in roughened, dull surfaces which might perhaps better be described as abraded, but following other microwear researchers, these are also included under the rubric of "polish". Characteristics of the polishes recorded at 200 x to 300 x include: brightness, surface texture, location, distribution, and the presence and orientation of trend lines.

The brightness of a polish refers to the degree it reflects light at high magnifications and was subjectively described as dull, moderately bright, bright or very bright. In comparison with an unmodified basalt surface dull polish reflects considerably less light, moderately bright polish reflects slightly less to an

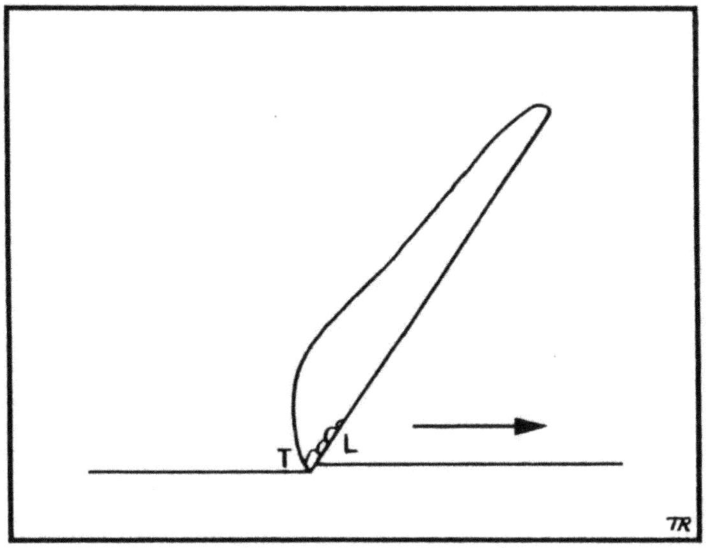

Figure 27. Leading (L) and trailing (T) edge-aspects of a tool used with a scraping action. The arrow indicates the direction of tool movement.

equivalent amount of light, bright polish reflects the same amount of light to more light, and very bright polish is much more reflective.

Surface texture of a polish is described as grainy, smooth, pitted or a combination of several. A very rough surface was described as "grainy" due to its granular appearance. Pits were exclusively present in the surface of smooth polishes and were described as to their presence or absence, relative frequency, size and shape.

Linear features or "trend lines" visible in the surface of the polish can be distinguished from striations, which are essentially scratches in the surface of a polish (see also Hayden 1979b:213; Keeley 1977:113-114). Trend lines can be very wide and involve considerable variations in surface topography, sometimes having the appearance of broad troughs. In many cases, trend lines are merely suggestions of linear features in the polishes only visible by moving the objective slightly in and out of focus. These linear features invariably indicate whether tools were used with an action transverse or longitudinal to the edge. Trend lines were recorded as to their characteristics and, more importantly, their orientation vis-a-vis the edge.

6.5.5 *Striations*

The importance of striations for microwear research was forcefully pointed out by Semenov (1964), however, this wear class is not commonly found and striations are usually only clearly observable at high magnifications. In this study striations were generally infrequently encountered, but since they so clearly reflect tool action considerable effort was taken to search for them to determine if they are correlated with specific tool actions, contact materials or conditions (i.e. grittiness). In many cases, striations were only suggested at 300 x, and magnifications of 600 x or higher were necessary to clearly view them. Following other researchers (e.g. Semenov 1964; Keeley 1980), the relative frequency of striations, location, length, width, and depth, relation to other types of microwear present, and orientation to the edge were recorded (Figure 28). Short striations are up to 20 µm long, medium striations are 21 µm to 50 µm in length, and long striations are greater than 50 µm in length.

6.6 *Natural "Wear"*

Since the present study is concerned with wear traces on experimental tools, the possibility of confusing non-use alterations of stone tools with traces of use-wear is not of immediate concern. However, since the ultimate goal of this research is to provide a means for interpreting the function of prehistoric tools, the matter should be addressed. The experimental program is primarily directed at providing a means for interpreting tools less than about 3000-4000 years old, and secondarily, tools to about 10,000 years. This is not enough of a time span to

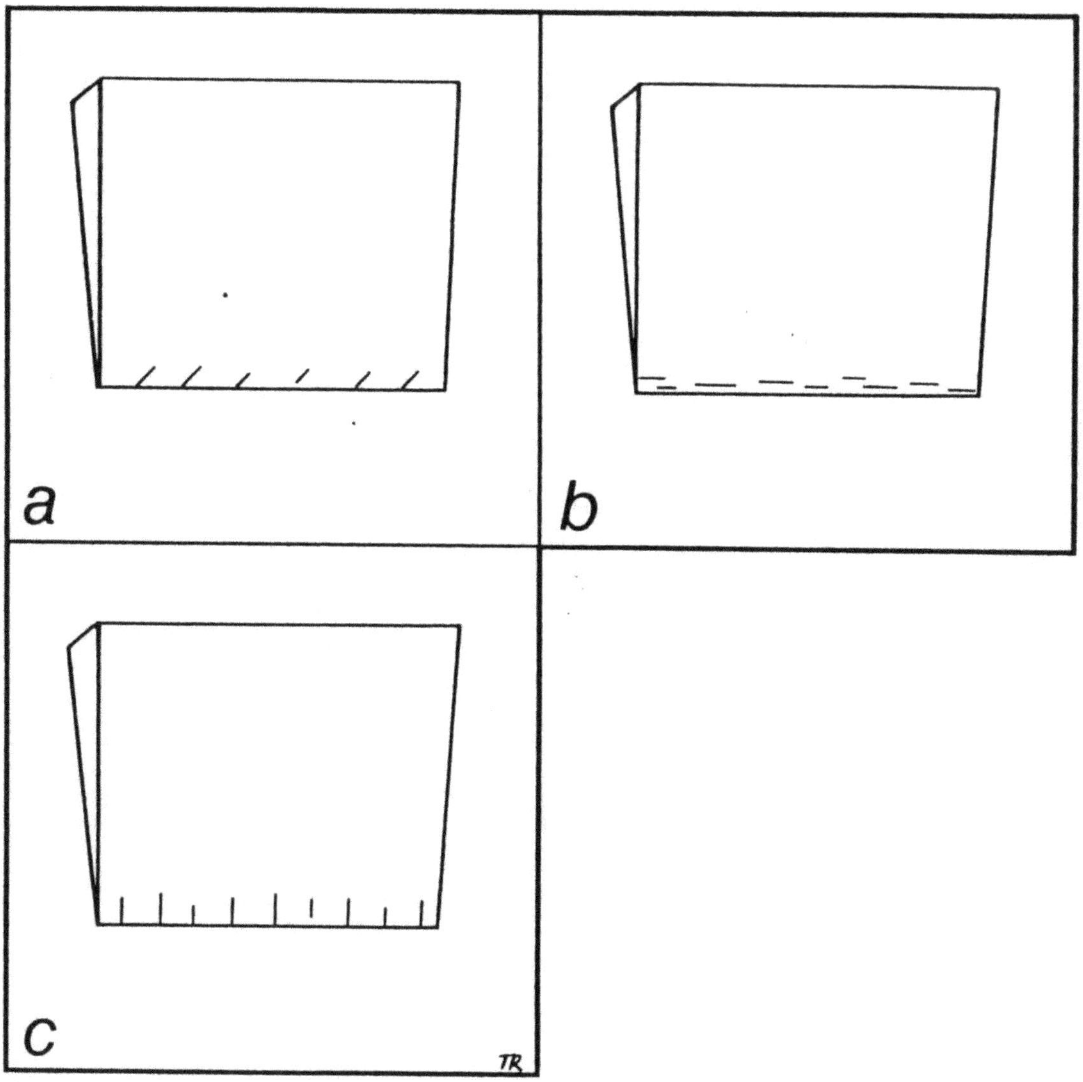

Figure 28. Orientation of striations in relation to a tool edge.
a. oblique, b. parallel, c. right-angle.

allow natural alterations such as patination, wind or water abrasion to be epidemic, as at Old World Palaeolithic sites. In my experience with late prehistoric Canadian Plateau assemblages, natural alterations are distinct and easily recognizable and are more likely to obscure use-wear rather than be confused with it, as is the case with European Palaeolithic tools (Keeley 1980:28-35).

CHAPTER 7

MICROWEAR OBSERVED ON EXPERIMENTAL BASALT TOOLS

7.1 *Introduction*

Detailed observations of microwear on 113 experimental Cache Creek basalt tools are presented in this chapter. Textual descriptions of microwear are supplemented by tabulated raw micorwear data (Appendix A). Light microscope and scanning electron microscope photomicrographs of the microwear (Appendixes B and C) and illustrations of the experimental tools (Appendix D) are also provided.

The chapter is divided into seven sections corresponding to major contact material classes and further subdivided on the basis of more specific contact materials and/or tool actions. The characteristics of each of the four major types of microwear (microflake scars, edge rounding, polish, striations) are described for each activity. Polish and striations observed with the aid of a scanning electron microscope (SEM) are described seperately from microwear observed with light microscopes.

7.2 *Plants (N=5)*

7.2.1 *Green Plants*

Two vitreous basalt tools were used on green reeds (*Juncus sp.*) and one on green wild grass (*Calamagrostis inexpansa*) intermixed with lesser amounts of sedge (*Carex atherodes*) (Tools 12-9,12-12,134-11; Figure 126a,d,e). The reeds stood in a shallow pond, while the grass and sedge were located on the edge of a dry slough. A handful of stems were held in one hand and cut near the bottom with a slicing motion. Only light to medium pressure was necessary to slice through a handful of stems. Duration of tool use ranged from 6 to 23 minutes (200-1750 strokes), and even the most heavily used tool was still usable for further cutting.

Microflake Scars: Snap (41.8%) and feather (40.0%) scars are the two most common microflake termination types. The majority of scars are small in size (83.6%). The ventral to dorsal scar ratio is 1.6:1, and the mean number of scars per tool is 18.3. Scars are rare and scattered on the two tools used on reeds, while they are contiguous and overlapping, as well as scattered, on the tool used on grass and sedge.

Edge Rounding: Edges are unmodified to slightly rounded.

Polish: Polish is bright to very bright, smooth and pitted (Figures 34,35). Pits are common and small to large in size. Polish tends to be nearly continuous along the edges of the two more heavily used tools (12-12,134-11) and scattered along the edge of the lightly used one (12-9). On the edge aspects, the polish is most heavily developed on high areas of the microtopography and in some areas these domed high points are joined by ridges of polish, forming networks of polish. On the heavily used tools the polish extends a considerable distance from the edge.

Polish (SEM): Polish is extensive and continuous along both edges of tool 12-12 and extends a maximum of 1 mm to 2.5 mm onto either edge aspect (Figures 2,73). The most highly developed polish usually extends no more than .5 to .75 mm from the edge. The polish is very smooth in texture, but pitted in even the most highly polished areas (Figure 74).

Striations: Striations are present on areas of high polish development on all three tools, but are only visible at high magnifications. They are short, narrow and the vast majority are oriented parallel or oblique to the edge, with a few oriented at or near right angles.

Striations (SEM): Striations are common on the edge segment of tool 12-12 (Figure 74). These striations are 3 µm to 15 µm in length and are principally oriented parallel to the edge, although a few are oriented oblique to the edge or at high angles to it.

Comparisons: There is only one other recorded example of basalt tools being used to cut green plants. Odell and Odell-Vereecken (1980:114-115) used a retouched Cache Creek basalt tool to cut common lawn grass in their "blind test" experiments. The only wear mentioned for this tool is "unmistakable corn gloss" (highly reflective polish), which is also present on the plant slicing tools of the present study.

7.2.2 *Dry Plants*

Two vitreous basalt tools (Tools 43-2,161-11; Figure 126b,c) were used for 3 to 19 minutes (150-700 strokes) to slice dry, yellow cattail (*Typha latifolia*) stems standing in a dry swamp. A single stem was held in one hand and cut just below the head with a slicing motion. Only moderate applied pressure was necessary to cut through a stem. Even the most heavily used tool was only slightly dulled at the end of the experiment.

Microflake Scars: Feather-terminated microflake scars are the most common type (59.1%), although a substantial proportion of scars are step-terminated (30.4%). As with tools used on green plants, most scars are small in size (81.7%). The ventral to dorsal microflake scar ratio is 1.2:1, and the mean number of scars per tool is 57.5. Scars are scattered in small, overlapping groups.

Edge Rounding: Edges are unmodified to slightly rounded.

Polish: Polish is bright to very bright, smooth textured, with small to large, circular to oval pits (Figure 36). The polish has a "flowing appearance" and is present on the edge and both edge aspects of the tools. On the heavily used tool 161-11 the polish is almost continuous on the edge, but scattered on the edge aspects near the edge. It is concentrated on the ends of snap scars and other projecting parts of the edge. On the lightly used tool 43-2, polish is present at a few loci on the edge and only slight indications are found on either edge aspect.

Striations: A few short striations oriented parallel or oblique to the edge are present on tool 161-11.

7.3 *Hide (N=22)*

7.3.1 *Hide Dehairing*

Eight vitreous and non-vitreous basalt tools (Tools 1-3,12-13,17-14,22-7,23-3,27-16,39-1,44-4; Figures 127a-f, 128a,b) were used to dehair a relatively fresh fallow deer hide (Figure 20). It was worked from 8 to 12 days after removal from the deer, and was somewhat stiff and dry throughout the experiments. Tools were used for 1 1/2 minutes to 47 minutes (200-5200 strokes). All but two tools (12-13,39-1) were unhafted; a few unhafted tools were wrapped in a small piece of rawhide to protect the hand. The flesh side of the hide was rinsed with running water which may have removed some grit from the hair side, however, dried mud later accidently accumulated on the hide and added grit. The hide was placed on plastic covered cardboard on the lab floor and "staked out" by tying corners of the hide to table legs (Figure 20). All of the tools proved effective for dehairing hides, although the hafted tools were easier to use than the unhafted ones.

Microflake Scars: Most microflake scars are step-terminated (59.2%) (Figure 75) although feather-terminated scars are also common (30.0%). Scars are mainly small (85.5%) and the mean number of scars per tool is 47.5. There are very few flake scars on the ventral edge aspect and half of the tools have none. The ventral to dorsal flake scar ratio is .14:1. Scars on the ventral aspect are primarily scattered, while those on the dorsal edge aspect are mainly contiguous and overlapping.

Edge Rounding: Edges tend to be rounded (Figures 39,40,75,76), with the rounding or facetting extending predominantly onto the dorsal (trailing) edge aspect.

Polish: The polish is uniformly dull (less reflective than the surrounding, unworn basalt surface) and grainy-textured (Figures 37-40). It is usually extensive and continuous along the edge and dorsal edge-aspect, but is present only in the immediate vicinity of the edge on the ventral edge-aspect. Rounded parts of the edge and scar ridges always have polish present. The boundary between polished and unpolished surfaces is usually relatively abrupt. At 600 x the polished surface appears to be covered with tiny lines, creating a "contour map" texture. Trend lines are present in the surface of the polish indicating the direction of tool movement (Figure 40). These are shallow, broad and long and run at approximate right angles to the edge. Polish is not distinguishable on two non-vitreous basalt tools (1-3,23-3). Both tools were used for moderately long periods and the edges are extensively rounded and appear "smoothed" in comparison with unworn surfaces.

Polish (SEM): The polish on tool 17-14 is rough-textured, and it is apparent at magnifications of 350 x and higher that the polish is not continuous, but comprised of numerous small, overlapping polished "plates" with gaps of various sizes between them (Figures 77-79). At magnifications above 1000 x it is apparent that the polished plates are themselves smooth-textured (Figure 79). The edge of the tool and the dorsal aspect have been extensively rounded by an apparently abrasive removal of the basalt. The polish is present on the rounded surface and gradually merges with the unpolished surface. The highly polished surface extends a maximum of .6 mm from the edge, while the maximum extent of polish extends up to 1 mm from the edge on the dorsal aspect. On the ventral aspect the polish is restricted to the close vicinity of the edge.

Striations: Striations are rare and located only on polished areas (Figures 37,38). They are small to medium in length and are primarily oriented at right angles to the edge, although a few are oblique to it.

Striations (SEM): Striations are not common on tool 17-14, and those present are oriented at or near right angles to the edge (Figures 77-79). They are relatively deep and wide, although the length varies considerably, with a few being exceptionally long. The largest striation is 70 µm long and 3 µm wide, while the shortest is 22 µm long. At very high magnifications, several parallel scratches can be seen to make up the 70 µm long striation (Figure 79). The striations first become visible at 200 x and are only present on the polished surface.

7.3.2 *Hide Defleshing (Scraping)*

Seven vitreous and three non-vitreous basalt tools (Tools 3-3, 3-4, 22-12, 27-17, 31-2, 44-6, 46-4, 80-4, 80-5, 80-6; Figures 128 e,f, 129 a,b,e,f, 130 c-f) were used to deflesh one fallow deer and one mule deer hide (Figure 21). One day old mule deer hide was partially defleshed by scraping with one unretouched (80-6) and two retouched tools (80-4,80-5). Tools were used from 30 seconds to 5 minutes on the hide, which was placed on a cement floor in a garage, hair side down. Tools tended to slip and were generally ineffective.

A four day old fallow deer hide was placed hair side down on plastic covered cardboard lying on the floor and was stretched out by tying edges of the hide to table legs. The hide was defleshed successfully and completely with seven tools used for 5 to 58 minutes (250-3200 strokes). Five of the tools were hafted (3-3,22-12,27-17,31-2,46-4) (Figure 21), while the rest were hand-held. The hafted tools were easier to use than unhafted ones, and vitreous basalt tools were more effective than non-vitreous ones.

Microflake Scars: The majority of microflake scars on unretouched tools are feather-terminated (57.0%), with step scars being the next most common type (23.3%). Most of the scars are small (83.9%). The ratio of ventral to dorsal microflake scars is .7:1, which is rather high for a scraping activity, but it is suspected that pronounced edge rounding and facetting has obliterated many of the scars on the dorsal edge aspect. It is suggested that the microscarring occurred first, followed by attrition of the edge resulting in rounding and polishing, which not only obliterated most flake scars, but eliminated suitable platforms for the detachment of more microflakes. Scars are mainly scattered, occuring singly or in small clusters of overlapping scars. The mean number of flake scars per tool is a low 24.1, however, the three non-vitreous basalt tools have very few microflake scars which reduces the average number of flakes on all tools and also affects the ventral to dorsal flake scar ratio. For vitreous basalt tools the average number of flake scars is 33.2, while it is only 9 per tool for non-vitreous basalt tools. The ventral to dorsal flake scar ratio is .6:1 for vitreous basalt tools and .9:1 for non-vitreous tools.

Edge Rounding: Edges become rounded with only slight use, and heavy use can cause the edge to become very rounded. The rounding extends predominantly onto the dorsal (trailing) aspect of the tool edges (Figures 80,81).

Polish: Hide defleshing polish is dull and grainy-textured in appearance (Figures 40-43). The boundary between a polished and an unused surface is quite sharp and distinctive. Polish is usually continuous and very well developed on the edge itself, which is usually rounded and thus quite broad. Trend lines in the polish run at approximate right angles to the edge.

Polish (SEM): Rough textured, "platey" polish (Figure 83) is continuous along the edge of tool 27-17 and both edge aspects, although penetration is very restricted on the ventral aspect. A large portion of the dorsal face is polished since the tool was tilted so much during use that this area frequently contacted the hide. However, highly developed polish is restricted to the edge and the dorsal aspect, and usually extends .3 to .75 mm from the edge and reaches a maximum of 1.3 mm. Trend lines in the polish can be observed at relatively low magnifications. These are oriented at or near right angles to the edge and tend to converge towards the center of the used edge.

Striations: Striations are extremely rare and were only noted on two tools (46-4,27-17). Single striations oriented at right angles to the edge are present on both tools, which are among the most heavily used.

Striations (SEM): Striations are very rare and are generally only visible at high to very high magnifications on tool 27-17 (Figure 83). The usual length is around 15 µm, and they are oriented at or near right angles to the edge. One very large gouge-like striation is visible without magnification and measures 2 mm in length and up to 70 µm wide (Figures 80-82). It begins in the highly polished area and arcs onto the unpolished surface.

Comparisons: There are few other experiments involving hide scraping to compare wear patterns with. Kamminga (1978: 492, 505) used altered and andesitic basalt tools for defleshing fresh kangaroo hide and scraping dry kangaroo hide, with sand added in a few cases. The tool edges were smoothed and rounded, one had polish, and several had microflakes and striations. The wear described by Kamminga is similar in general characteristics to that observed here for hide defleshing and in section 7.3.1 for hide dehairing.

7.3.3 Hide Defleshing (Slicing)

Two vitreous basalt tools (Tools 80-7,80-8; Figure 129c,d) were used to cut fat from fresh mule deer hide which was laid out on the cement floor of a garage. The experiments were an attempt to remove adhering flesh by slicing, as scraping with retouched, unhafted tools had proved ineffective. Slicing did not work much better and the tools were only used for 30 seconds and 1 minute respectively.

Microflake Scars: Half of the microflake scars present are hinge-terminated (50.0%), and a substantial proportion are snap-terminated (27.8%). Low percentages of feather- (16.2%) and step-terminated scars (5.9%) are also present. Almost all of the scars are small (98.5%). The ventral to dorsal scar ratio is .8:1, although the individual tool ratios vary considerably. The mean number of flakes per tool is 34. Flake scars are mainly scattered singly or in small groups, occasionally being in overlapping and superposed configurations.

Edge Rounding: Absent.

Polish: Polish is dull, and grainy-textured in appearance. It is restricted to the edges, where it occurs in long, discontinuous patches.

Striations: Absent.

7.3.4 Dry Hide Sawing

Two vitreous basalt tools (Tools 12-17,134-9; Figure 128c,d) were used to saw through dry, stiff fallow deer hide (13 days old) which had been both dehaired and defleshed. Tools were used for 4 to 18 minutes (400-1200 strokes). Medium to extremely heavy pressure had to be applied to saw through the hide, and the tools were noticably dulled during use.

Microflake Scars: Almost half of the microflake scars are step-terminated (47.0%), although-feather terminated scars are also common (36.8%). Scars are mainly (83.1%) small in size. The mean ventral to dorsal flake scar ratio is .7:1. Scars are primarily contiguous and overlapping, with small scars superposed over larger ones.

Edge Rounding: Edges are slightly rounded (Figure 84).

Polish: Polish is present on the edge and both aspects of the two tools. It is dull and rough textured, with a grainy appearance (Figures 44,45). The polish is continuous along the entire used portion of the tool edges and is generally restricted to small patches on high points of the edge-aspects near the edge.

Polish (SEM): Polish on tool 134-9 is restricted to the edge and along scar ridges and the backs of step scars on both aspects in close proximity to the edge (Figures 84,85). However, weakly developed polish, which apppears as smoothing of the tool surface, extends much farther back from the edge. Highly developed polish extends a maximum of 50 µm from the edge and usually much less, while the smoothing extends as much as 150 µm from the edge. The polished surface is smooth and unpitted, but in places tends towards the "platey" texture observed on hide scraping tools. Unlike the "platey" polish on defleshing and dehairing tools, this polish is only visible at very high magnifications (Figure 86).

Striations: Striations are not observable with the aid of an incident light microscope on either of the tools.

Striations (SEM): Striations appear to be common on polished areas of tool 134-9, and are oriented parallel or oblique to the tool edge (Figures 85,86). They are narrow and shallow and vary in length from 2 µm to 12 µm.

Comparisons: Schutt (1982: Appendix 2) used one basalt tool for sawing commercially tanned suede. No polish was observed, but edge rounding was present as well as bifacial snap-terminated scars. Kamminga (1978:524-525) used altered and andesitic basalt tools for cutting kangaroo hide and observed edge rounding, blunting, snap scars and other microflake scars on them. Although Schutt and Kamminga did not observe polish or striations, the wear they did observe is broadly similar to that on the hide sawing tools studied here.

7.4 Bone (N=10)

7.4.1 Unsoaked Bone Sawing

Four vitreous and non-vitreous basalt tools (Tools 21-6, 26-8,26-9,46-2; Figures 123b, 124b,d, 125f) were used to saw fresh, unsoaked fallow deer bone for 1 ½ to 4 minutes (200-400 strokes). The tools were held at approximate right angles to the surface of a long bone and were vigorously drawn back and forth while attempting to saw through the shaft. After 4 minutes of sawing with heavy applied pressure only a narrow, shallow groove was sawn into the bone. Tool edges were massively damaged after very little use.

Microflake Scars: There are moderately high percentages of hinge (38.8%) and step (35.9%) scars, while feather scars are relatively rare (12.0%) (Figure 80). Snap-terminated scars are uncommon (13.4%), but are significant wear features due to their large or very large size which lends the edges a serrated appearance. Most scars (69.4%) are small in size. The ventral to dorsal scar ratio is 1.9:1, and the mean number of scars per tool is 52.3. Scars are contiguous and overlapping, although there are areas lacking scars due to the alternating nature of the use-retouch.

Edge Rounding: The edges are blunted to very slightly rounded in places.

Polish: Bright, smooth-textured and pitted polish is rare and restricted to scattered high points on the edge and edge-aspects (Figures 41-44). Bright, grainy-textured polish is much more common, and dull, grainy polish is also present.

Polish (SEM): Absent (46-2).

Striations: Possible striations oriented at oblique angles to the edge of tool 46-2 were observed at high magnifications (Figure 44).

Striations (SEM): Absent (46-2).

Comparisons: Kamminga (1978) sawed bone with altered, andesitic and olivine basalt tools. The most prominent wear consisted of large snap scars, with smaller scars within them; edge rounding, blunting and smoothing were also present (Kamminga 1978:151-156,544-545). Montgomery (1978:274) sawed fresh cow bone with one glassy basalt tool, which resulted in a jagged edge with bifacial microflake scarring. Schutt (1982:Appendix 2) sawed fresh bone with two basalt tools and observed edge rounding, as well as bifacial scarring primarily consisting of feather scars, with some step and snap scars. The wear produced in all three of these bone sawing experiments is more or less similar to the wear described in the present study, with that described by Kamminga being most similar. None of these researchers reported polish, which the present experiments indicate is only visible at magnifications of 200-300 x with a light microscope.

7.4.2 Soaked Bone Sawing

Two basalt tools (Tools 11-6,18-19; Figure 124a,f), one vitreous and one non-vitreous, were used to saw soaked fallow deer bone for approximately 1 minute each (200 strokes). The tools were used with heavy applied pressure while attempting to saw through a long bone shaft, but both tools became totally dulled during use. Soaking in water for 13 days did not make the bone any easier to work than fresh, unsoaked bone.

Microflake Scars: Most microflake scars (59.3%) are step-terminated, although a significant percentage (23.1%) are feather-terminated (Figure 81). Three-quarters (75.0%) of the scars are small. Very large snap scars are not numerous but are very prominent on the edge, giving it a serrated appearance. The ventral to dorsal flake scar ratio is 1:1 and the average number of flakes per tool is 46. Microflake scars are contiguous and overlapping, although some areas are devoid of scars, due to the alternating nature of the retouch.

Edge Rounding: Edges are blunted and very slightly rounded in places.

Polish: On tool 18-19 the polish is dull and rough textured, similar to polish on hide-scraping tools, but slightly brighter. The polish on tool 11-6 has tiny bright to very bright smooth spots surrounded by grainy, dull polish similar to that on 18-19. Polish on both tools is rare and scattered, being mainly found on the edges, but occasionally extending slightly onto either aspect.

Polish (SEM): Absent (18-19).

Striations: Absent.

Striations (SEM): Possible striations are visible at one locus on tool 80-19 with very high magnifications (above 3000 x). They are oriented parallel or oblique to the edge. Most of the observed striations are about 8 μm long, with the largest measuring 12 μm.

7.4.3 *Unsoaked Bone Scraping*

A single vitreous basalt tool (Tool 22-20; Figure 123a) was used to scrape an unsoaked, fresh fallow deer long bone. The bone was scraped for about 2 minutes (70 strokes) with heavy applied pressure, which had little effect on the bone while resulting in pronounced dulling of the tool.

Microflake Scars: As with sawing fresh, unsoaked bone there is no clearly dominant microflake termination type, although hinge scars are also the most common type (37.8%). Both step- (32.7%) and feather-terminated (28.6%) scars are strongly represented. Almost all of the scars (92.9%) are small in size. The ventral to dorsal scar ratio is 1.3:1 which is unusually high for scraping and is entirely due to the dorsal edge-aspect leading and the ventral aspect trailing during use, as necessitated by the morphology of the tool. There are 98 scars on this tool, and these are contiguous, overlapping and superposed.

Edge Rounding: The edge is slightly rounded in places.

Polish: The scattered polished loci consist of small, bright to very bright, smooth core areas surrounded by duller, grainy polish. Often, the dull, grainy polish occurs without the bright polish. The smooth, bright polish is mainly found on high points or projecting parts of the tool, primarily on the edge, but also on both aspects near the edge.

Striations: Absent.

Comparisons: Montgomery (1978:273, Figure 67) scraped fresh bone with tools having low and high spine-plane angles. She observed unifacial (dorsal), small, overlapping scars on the low-angled tool. Scars were more rare and smaller on the steep-edged tool, with a few on the ventral edge aspect. Schutt (1982:Appendix 2) scraped fresh bone with an unretouched basalt tool. Microscarring was bifacial, although the majority of scars were present on the ventral aspect. Most were step terminated and bifacial edge rounding was also present. The wear observed by Schutt is most similar to the wear produced in the present series of experiments, especially with the majority of scars being removed from the leading (ventral) edge of the scraping tools. Neither Schutt nor Montgomery observed polish, most likely due to the low magnifications they employed.

7.4.4 *Soaked Bone Scraping*

One vitreous basalt tool (Tool 17-13; Figure 124e) was used for scraping fallow deer bone which had soaked in water for 13 days. The tool was used for about 1 1/2 minutes (100 strokes) and was found to be useful for removing tissue remaining on the bone and smoothing the surface slightly, but it was virtually useless for substantially thinning the bone. Soaking does not seem to have made the bone easier to work.

Microflake Scars: The predominant microflake scar type (83.6%) is stepped (Figures 83,84) and most scars (73.8%) are small in size. The ventral to dorsal scar ratio is .07:1 and there are 61 scars on the tool. On the ventral edge-aspect the few scars present are overlapping and clustered in one group. On the dorsal edge-aspect scars are contiguous, overlapping and often superposed (Figures 83,84).

Edge Rounding: Absent.

Polish: Polish is nearly continuous along the edge but is restricted to scattered, small patches on the dorsal edge-aspect and is absent from the ventral edge-aspect. The polish is grainy-textured, and dull with occasional bright, smooth spots (Figures 45-46). One larger, bright, smooth area was present on the dorsal edge-aspect (Figure 47).

Polish (SEM): Some smoothing is visible near the edge at very high magnifications, but no actual polish is discernable.

Striations: Absent.

Striations (SEM): Absent.

7.4.5 *Whittling Bone*

One vitreous basalt tool (Tool 63-1; Figure 124c) was used in an attempt to thin a fresh, unsoaked fallow deer long bone by whittling. The tool was used for 1 minute (60 strokes) and proved to be totally useless for accomplishing the intended task.

Microflake Scars: Microflake scars are predominantly (67.7%) hinge-terminated, with much lower percentages (16.2%) of step, feather (8.8%) and snap (7.4%) scars. Most scars are small in size (83.8%), and the ventral to dorsal scar ratio is.4:1. The 68 scars on this tool are scattered singly or in small overlapping groups on the ventral aspect and are contiguous, overlapping and superposed on the dorsal edge aspect.

Edge Rounding: The edge is blunted with slight rounding in places.

Polish: Polish is present on the edge and the dorsal edge-aspect. It is mainly dull and grainy, with the occasional bright, smooth spot.

Striations: Absent.

Comparisons: Only one other researcher has attempted to whittle bone with a basalt tool. Schutt (1982:Appendix 2) whittled fresh bone and observed unifacial edge rounding as well as bifacial microscarring consisting of step and feather scars. The wear on the tool used in the present experiment differs in that hinge scars are predominant, there are many more scars, and polish is present.

7.4.6 *Graving Bone*

One vitreous basalt tool (Tool 138-19; Figure 125e) was used to engrave a fallow deer longbone. The tool had a spurred tip which rapidly broke down during use, although it stabilized and finally dulled to near uselessness after 12 minutes of graving (600 strokes). Deep, parallel grooves were successfully engraved as in the "groove and splinter" technique.

Edge Rounding: Edges and ridges near the tip are well rounded.

Polish: Polish is continuous and concentrated near the tip. It is very bright, smooth-pitted and lends the surface a metallic appearance (Figure 48). In one area numerous, shallow, comet-shaped pits are present. These parallel the direction of tool motion. Also present is grainy-textured, dull to bright, polish.

Striations: Absent.

7.5 *Antler (N=17)*

7.5.1 *Sawing Unsoaked Antler*

Fresh fallow deer (6 weeks old) and old mule deer antler (4 years old) are grouped together here since there does not seem to be significant differences in the wear patterns on vitreous basalt tools produced by working either of these materials. Three vitreous basalt tools (Tools 43-3,134-22,141-7; Figure 125a, 126d, 128b) were used for .5 to 2 minutes (200-400 strokes) with sawing actions across the antler beams with the intent of cutting through them. Although heavy pressure was applied, the tools were relatively ineffective at sawing antler, and managed only to cut shallow, narrow grooves while rapidly becoming dulled.

Microflake Scars: Step-terminated scars are most common (47.8%) (Figures 85,86), although feather-terminated scars are also abundant (39.1%). Considerably more than half (63.9%) of the scars are small in size and the ratio of ventral to dorsal microflake scars is 1.4:1. The mean number of microflake scars per tool is 84. Microflake scars are generally contiguous and overlapping with smaller scars often superposed over larger ones (Figures 85,86).

Edge Rounding: Absent.

Polish: Polish is mainly present along the tool edges, and extends slightly onto both edge-aspects. Small patches of bright, but grainy-textured polish is located on the highest parts of the microtopography. It is usually surrounded by more extensive dull, grainy-textured polish.

Polish (SEM): Absent (134-22).

Striations: Absent.

Striations (SEM): Absent (134-22).

Comparisons: Montgomery (1978:Fig. 66) used one glassy basalt tool to saw dry deer antler, which resulted in wear consisting of "...overlapping broad, crescent shaped scars" (Montgomery 1978:274). In addition, she also observed that working antler in general produced "edge crushing and rounding", while sawing antler generally produced "a jagged or 'scalloped' edge consisting of wide shallow scars on both faces" (Montgomery 1978:274-275). The wear observed by Montgomery is similar to the wear observed in the present study, and although edge rounding was not observed on sawing tools used on dry antler, it was present on tools used to saw soaked antler.

7.5.2 *Sawing Soaked Antler*

Five vitreous (Figures 127a-c, 128a,c) and two non-vitreous (Figures 126b, 127f) basalt tools were used to saw soaked antler (Tools 11-2,11-12,17-11,27-3,36-7,18-17,141-12). Four year old mule deer antler and fresh fallow deer antler were worked after being soaked in water for 9 to 13 days. Tools were used with a sawing motion while attempting to cut through antler beams for 1 to 2 minutes (200-400 strokes), with heavy applied pressure. Narrow, shallow grooves resulted from the sawing, but the tool edges were virtually destroyed. Soaking definitely softened the antler, especially the fresh antler. Also noteworthy is the fact that antler is noticably softer after 13 days of soaking than after only 9 days of soaking.

Microflake Scars: No one type of microflake termination predominates, although feather scars are the most common (32.4%). Step-terminated scars are the next most common type (27.7%), followed by snap (22.8%) and hinge (17.2%) scars. The main difference between working soaked old antler and soaked fresh antler is in the mean number of microflakes per tool, and even this is slight. Most flakes (74.8%) are small in size, but deep, and a substantial percentage (15.4%) are medium-sized. Although relatively rare, the large to very large snap scars on each tool dominate the edge (Figure 87) and in several cases smaller scars are found within them. The ventral to dorsal flake ratio is 1.2:1. Flakes are primarily contiguous and overlapping with small scars superposed over larger ones.

Edge Rounding: The edges are moderately rounded or blunted with slight rounding.

Polish: Polish is nearly continuous on the tool edges, but is concentrated on the high points of both edge-aspects. It is grainy-textured and dull to bright. It is not possible to clearly distinguish polish on the two non-vitreous basalt tools (11-2,11-12).

Polish (SEM): Absent (17-11).

Striations: Absent.

Striations (SEM): Absent (17-11).

7.5.3 *Scraping Unsoaked Antler*

Dry, unsoaked, four year old mule deer and six week old fallow deer antlers were scraped by vitreous basalt tools (Tools 22-11,138-13; Figures 125b, 126a) with heavy applied pressure for about 1 minute (100-115 strokes). Scraping accomplished a smoothing of the antler surface, but substantial thinning was not possible. Both tools were severely dulled by 80 to 100 strokes.

Microflake Scars: Most microflake scars (69.3%) are step-terminated (Figure 88). Over half of the scars (53.3%) are small, but a substantial percentage (20.4%) are medium-sized or large (16.8%). The mean number of scars per tool is 68.5 and the ratio of ventral to dorsal microflake scars is .2:1. Scars are scattered to absent on ventral (leading) edge aspects, while they are contiguous and overlapping with superposition of small scars within larger ones on dorsal (trailing) aspects (Figure 88).

Edge Rounding: There is slight edge rounding.

Polish: Polish is bright and grainy-textured. It is present primarily on the edges and dorsal aspects, especially along the back of step scars.

Polish (SEM): Absent (138-13).

Striations: Absent.

Striations (SEM): Absent (138-13).

Comparisons: Montgomery (1978:Figures 65,66) used three tools to scrape dry deer antler. She observed edge crushing, rounding and large overlapping flake scars (mainly on the dorsal edge-aspect) (Montgomery 1978:273). With the exception of polish, this wear is similar to that observed on unsoaked and soaked antler scrapers in the present study.

7.5.4 Scraping Soaked Antler

Three vitreous basalt tools (Tools 141-14,161-18,161-19; Figure 125d, 126c, 127e) were used to scrape four year old mule deer antler and fresh fallow deer antler which had soaked in water for nine days prior to the experiments. Tools were drawn towards the experimenter with a scraping motion for 1 to 2 minutes (100-200 strokes) with heavy applied pressure. They were effective at smoothing the antler and slightly thinning it, but were not suited for significant alterations.

Microflake Scars: The most common (45.8%) flake termination type present is the step scar, although hinge (30.8%) and feather scars (22.1%) are also well represented (Figure 89). Three-quarters (74.6%) of the microflake scars are small in size. The mean number of scars per tool is 99.7 and the ventral to dorsal microflake ratio is .4:1. On the ventral edge-aspect, scars are mainly scattered, although they also occur in small groups of overlapping scars. On the dorsal edge aspect, microflake scars are contiguous and overlapping, with smaller scars often found within larger ones (Figure 89).

Edge Rounding: The edges are occasionally slightly rounded.

Polish: Polish is present on all three tools, especially on the edge and the dorsal edge-aspect. It is either bright and grainy-textured or dull and grainy. Patches of bright polish tend to be located on high parts of the microtopography. The bright polish is nearly continuous along the edge.

Polish (SEM): Absent (161-19).

Striations: Short, narrow striations are present in the polish on two tools (161-18, 161-19). These are not common, being present at single loci on or near the edges of both tools. All striations are oriented at or near 90 degrees to the tool edge.

Striations (SEM): Absent (161-19).

7.5.5 Boring Soaked Antler

A single vitreous basalt tool (Tool 134-6; Figure 132c) with a projecting spur was used to bore into nine month old fallow deer antler which had soaked in water for 20 days. Boring was undertaken with heavy pressure applied to the tool, which was held at approximately 90 degrees to the antler surface during use. It was employed for 15 minutes (1000 strokes), boring a substantial hole into the antler. Twenty days of soaking had softened the antler significantly more than soaking for 13 days.

Microflake Scars: Step-terminated scars are the most common (45.1%) scar type, but substantial percentages of feather (31.4%) and hinge (19.6%) scars are also present. The majority of the 51 scars on this tool are small (76.5%) and concentrated in groups of overlapping scars, with smaller ones superposed on larger scars. There are also occasional isolated single scars.

Edge Rounding: The tip and ridges near it are extensively rounded.

Polish: Polish is extensive near the tip of this tool and extends a considerable distance up ridges away from the tip. Much of the polish is grainy-textured, but very bright, although some dull polish is also present (Figure 54). Smooth-pitted polish is present on a few high points (Figures 54,56). There is a linear trending in the polish, consisting of broad, shallow grooves which run at or near right angles to the long axis of the tip (Figures 54,55).

Striations: Absent.

7.5.6 *Graving Soaked Antler*

A single vitreous basalt tool with a projecting spur (Tool 141-20; Figure 134d) was used for engraving nine month old fallow deer antler which had soaked in water for 20 days. The tool was used to score two parallel grooves in a flat part of the antler surface (for the eventual removal of a splinter as in the "groove and splinter technique"). The tool tip suffered considerable attrition, although it remained effective after 21 minutes of use (1000 strokes).

Microflake Scars: Half of the microflake scars (50.0%) are feather-terminated, although step-terminated scars are also common (37.5%). Nearly all (93.8%) of the 16 scars are small in size. They often overlap each other in small, scattered clusters.

Edge Rounding: The tip and associated ridges are extensively rounded.

Polish: Polish is present on the tip of the tool and on ridges near the tip. It is mainly rough-textured and dull to very bright (Figures 57-61), although bright, smooth-pitted areas are also present (Figures 60,61).

Striations: Striations are present in the bright, smooth-pitted polish (Figures 60,61). These are short and oriented parallel or oblique to the long axis of the graving tip.

7.6. *Fish Butchering (N=7)*

The five large trout (1.5-2.5 kg each) used in the fish butchering experiments were frozen for approximately 2 years prior to experimentation. The fish had been caught in northern Saskatchewan and were gutted prior to freezing.

7.6.1 *Slicing Whole Fish*

Three vitreous basalt tools (Tools 12-10,12-14,134-16; Figures 135e, 136b,d) were used to cut up fish, contacting flesh, bones, scales and skin in the process. The tools were used with a slicing motion for 5 to 10 minutes (200-500 strokes) with little loss of effectiveness. Applied pressure varied from light to heavy.

Microflake Scars: One tool (12-14), which was used for 5 minutes, has no microflake scars. The others have nearly even percentages of feather (48.5%) and step (42.7%) terminated microflake scars. The overwhelming majority of scars (93.2%) are small in size and the ventral to dorsal scar ratio is .9:1. Scars are mainly scattered singly or in small groups of overlapping scars. The mean number of scars per tool is 34.3.

Edge Rounding: Absent.

Polish: Polish is primarily dull and grainy-textured, although scattered patches of somewhat brighter, grainy polish are also present. The polish is mainly located on the tool edges with scattered occurrences on both edge-aspects.

Striations: Absent.

Comparisons: Odell and Odell-Vereecken (1980) used one Cache Creek basalt tool for sawing a fish, contacting flesh, skin and bone. The illustrated tool has a denticulated edge with overlapping feather and hinge terminated, microflake scars. In the present series of experiments, feather terminated scars are by far the most common type, followed by step scars, with very few hinge scars.

7.6.2 *Slicing Fish Flesh*

One non-vitreous (Figure 136a) and two vitreous basalt tools (Figure 135d,f) were used to cut trout flesh into fillets, occasionally contacting bone and skin (Tools 20-7,46-3,138-12). The tools were not noticably dulled after 2 to 10 minutes of slicing (200-1000 strokes) with light to medium applied pressure.

Microflake Scars: The non-vitreous basalt tool (20-7) was used for 4 minutes (500 strokes), but has no discernable microflake scars. Microflake scars on the two vitreous basalt tools are almost all (97.9%) small in size (Figure 97). The majority of scars (55.0%) are feather-terminated, although a substantial percentage (36.4%) are step-terminated. The ventral to dorsal flake scar ratio is 1.8:1 and the mean number of scars per tool is 47. Scars are mainly scattered singly or in small groups of overlapping scars, although on the heavily used tool 138-12 scars are overlapping and contiguous all along one edge-aspect.

Edge Rounding: Absent.

Polish: Only tool 46-3 has polish and this is dull and grainy, with the occasional brighter patch. It is present on the edge and both edge-aspects.

Polish (SEM): No true polish is present on the examined segment of tool 138-12. Instead, a discontinuous, slight smoothing of both edge aspects within 50 µm of the edge was observed.

Striations: Absent.

Striations (SEM): One possible medium-sized, deep striation, oriented roughly parallel to the edge was observed on the smoothed area of tool 138-12.

7.6.3 *Fish Scaling*

One fish scaling experiment was undertaken with a vitreous basalt tool (Tool 134-17; Figure 136c). The tool was most effective at a 40-50 degree angle to the contact surface and fish scales were removed with a slicing-whittling motion, somewhat like whittling wood. The tool was only slightly dulled after about 4 minutes of use (500 strokes).

Microflake Scars: Most microflake scars (80.0%) are feather terminated. All scars are small (to very small) in size and very shallow. They are bifacially distributed along the used edge, with a ventral to dorsal scar ratio of 3.5:1. Scars are rarely superposed and are contiguous and overlapping on the ventral edge aspect and scattered singly or in small groups on the dorsal aspect. There are 45 microflake scars on this tool.

Edge Rounding: Absent.

Polish: Absent.

Polish (SEM): Polish is present on the edge and both edge-aspects. Highly polished areas are patchy, although a light polish or smoothing is nearly continuous along both edge-aspects. The polish is intense or well-developed on high points and only becomes visible at magnifications above 200 x. It is very smooth, lacks true pits, and extends 15 µm to 40 µm from the edge. The light polish or smoothing extends a maximum of 80 µm from the edge.

Striations: Absent.

Striations (SEM): Absent.

7.7 *Deer Butchering (N=26)*

7.7.1 *Evisceration*

One vitreous basalt tool (Tool 12-3; Figure 140c) was used to eviscerate a fallow deer killed approximately one half-hour prior to the beginning of the experiment. It took 28 minutes (750 strokes) to cut through the hide, open the abdomen, cut out the guts and cut around the anus. Bone was nicked lightly only a few times. A slicing action was mainly employed, although sawing and piercing were also occasionally necessary. Light to medium pressure was applied to the tool. The tool was very well suited for the task, but somewhat dulled at its completion.

Microflake Scars: Step- (40.4%) and feather-terminated (37.5%) microflake scars are the most common scar types. The overwhelming majority (97.8%) of the microflake scars are small. The ventral to dorsal scar ratio is 1.8:1, with a total of 136 scars present on the tool. On the ventral edge-aspect, scars are predominantly contiguous and overlapping, with smaller scars superposed over larger ones. On the dorsal edge-aspect, the scars are primarily scattered in small groups of overlapping scars.

Edge Rounding: The edge is slightly rounded.

Polish: Polish is continuous on the tool edge, mainly bright and grainy, although some bright, smooth (but lined) polish is also present. On the edge-aspects, scattered polish is mainly dull to dull-bright and grainy (Figure 62), with less present on the dorsal than ventral.

Striations: A few possible short striations running parallel to the edge on the polish were observed at 400 x.

7.7.2 *Skinning*

Two freshly killed mule deer and one freshly killed fallow deer were skinned with five vitreous basalt tools (Tools 12-6,43-4,80-1,80-12,80-13; Figures 136e, 139c,f, 140e). Individual tools were used for 13 to 50 minutes with slicing and sawing actions. During the removal of the hide from the deer, hide was cut, as well as fat and flesh holding the hide to the carcass. Sand and silt were noted on the fallow deer hide and undoubtedly all of the hides had grit in the hair. Medium to heavy pressure was applied to the tools.

Microflake Scars: Slightly over half of the microflake scars (51.5%) are feather-terminated, and almost one-third (29.7%) are snap-terminated (Figure 100). The vast majority of scars (96.4%) are small in size, and several tools have only small scars. The ventral to dorsal scar ratio is 1.4:1, and the mean number of scars per tool is 66.4. Scars are primarily scattered, mainly in small clusters of several overlapping scars, with small scars often superposed over larger ones.

Edge Rounding: The edges are rounded or occasionally blunted.

Polish: Polish on the edges and edge-aspects is mainly dull and grainy, although bright, grainy polish and bright, smooth polishes are also present. Distribution of polish on the edges varies from only slight traces to nearly continuous stretches on different tools. It is generally extensive, but scattered, on both edge-aspects.

Polish (SEM): Polish is very extensive and visible at quite low magnifications (60 x) on tool 12-6 (Figure 100). Heavy polish is continuous along the very edge of the tool and extends onto the edge aspects along projecting spurs (Figures 100,101). This polish is domed, with a very smooth and unpitted surface (Figures 101,103). The heavy polish extends no more than .5 mm from the edge and usually less than .2 mm. A slightly smoothed or lightly polished area extends to about 1 mm from the edge.

Striations: Absent.

Striations (SEM): Striations are only clearly visible at high magnifications on tool 12-6 (Figure 103). They are relatively common and tend to be oriented parallel or oblique to the edge. Observed striations range from 4 µm to 16 µm long, and are narrow and shallow to deep.

Comparisons: Odell (1980) used several unretouched Cache Creek basalt tools to skin a raccoon and a porpoise. The wear he observed consisted primarily of bifacially distributed feather scars, producing a denticulated edge. In addition to the bifacial feather scars, striations and polish were observed in the present study.

7.7.3 Slicing Flesh

Freshly killed fallow deer flesh was cut with 4 vitreous and 3 non-vitreous basalt tools (Tools 1-4,11-9,22-17,27-5,35-3,36-8,45-7; Figures 139a,b,d,e, 140a, b, d). Tools were used with a slicing motion while removing meat from a deer haunch, cutting meat and fat from ribs, removing sinew, and cutting tenderloin into strips. Bone was only nicked a very few times by the tools. Individual tools were used between 1 and 34 minutes (50-2000 strokes). One non-vitreous basalt tool (1-4) was noticeably dulled after 9 minutes of use (700 strokes) (Figure 23), while a vitreous basalt tool (45-7) was only slightly dulled and still useful after 34 minutes (2000 strokes).

Microflake Scars: One of the non-vitreous basalt tools (1-4) has no discernable microflake scars and the others only have two scars each. These are all small, feather-terminated and located on the ventral edge-aspects. On the vitreous basalt tools, microflake scars are mainly step- (56.5%) and feather-terminated

(30.8%). Nearly all of the scars (97.9%) are small and shallow (Figure 104). The ventral to dorsal scar ratio is 1.2:1. The average number of scars per tool is 59.3. Scars are often scattered singly or in small clusters of overlapping scars. Only on the most heavily used tool (45-7) are the scars contiguous and overlapping on both edge aspects.

Edge Rounding: Edges are slightly rounded on the more heavily used tools.

Polish: On the vitreous basalt tools there are traces of dull, grainy polish, bright, grainy polish and a few spots of bright, smooth polish on the edges and edge-aspects (Figures 63-65). Smoothing of unpolished parts of the edge-aspects is also present. The non-vitreous tools have dull, grainy polish and bright, grainy polish on the edges and edge-aspects. The polish is nearly continuous on the most heavily used tool, but only traces of polish are present on the others.

Polish (SEM): Tool 45-7 has polish localized on prominences along both edge-aspects. This polish is smooth-textured, with a slight tendency to a "platy" texture around the peripheries which is reminiscent of polish on tools used to work hide (Figure 105). There is also a slight smoothing or light polish which extends .2 mm to .3 mm up from the edge.

Striations: Absent.

Striations (SEM): On tool 45-7, short striations oriented parallel or oblique to the edge are quite common on highly polished areas, and are also present on lightly polished areas.

Comparisons: Montgomery (1978:Figures 69,70) used an unretouched glassy basalt tool for sawing fresh meat, however, no wear was observed. This is almost certainly due to the very low magnification Montgomery used to study her tools. Odell (1980) used one Cache Creek basalt tool for cutting porpoise blubber and observed very small, scattered, bifacial, feather-terminated scars. This is similar to the bifacial, scattered, small microflake scars produced by slicing fresh flesh in the present experiments, although step scars are more common than feather scars.

7.7.4 *Dismembering (Slicing)*

Four vitreous basalt tools (Tools 12-15,22-10,22-15,45-6; Figures 136f, 138a,b,d) were used to dismember a fallow deer, which mainly involved cutting off legs and forelegs by slicing through joints. Hide, flesh, tendons, and ligaments were the primary materials contacted, although cartilage and bone were nicked accidently. Individual tools were used from 6 to 20 minutes (150-740 strokes). Pressure applied to the tools was mainly medium to heavy. One tool was noticeably dulled after 15 minutes of use (350 strokes), although another tool used for 20 minutes (740 strokes) was only slightly dulled and was still useful.

Microflake Scars: Step- (42.6%) and feather-terminated (38.5%) scars are most common. Scars are primarily small (90.8%) and shallow. The ventral to dorsal scar ratio is 1.1:1 and the mean number of scars per tool is 48.8. Scars are principally scattered singly or in small clusters of overlapping scars.

Edge Rounding: Edges are occasionally rounded or blunted.

Polish: The most heavily used tool (45-6) has nearly continuous bright grainy polish with scattered bright smooth spots along the edge and scattered along both edge-aspects. Also, smoothing of the edge-aspects beyond the limited polished areas is apparent. The other tools only have traces of dull, grainy polish.

Striations: Absent.

7.7.5 *Dismembering (Sawing-Slicing)*

Three tools (Tools 12-11,80-3,80-14; Figure 138c,e,f) were used to partially dismember two mule deer and one fallow deer. All of the tools used were made of vitreous basalt, and one (80-14) was unifacially retouched. Two of the tools (80-3,80-14) were used primarily with a sawing action, but some slicing also took place. One tool (12-11) was used primarily in a slicing motion, with a limited amount of sawing. The tools were used from 3 to 32 minutes (a maximum of 920 strokes), with medium to heavy applied pressure. They were specifically employed to remove the legs and cut off the forelegs of the deer (Figure 22), and in the process mainly contacted flesh, hide, and ligaments. Cartilage and bone were also occasionally contacted.

Microflake Scars: Microflake scars on the unretouched tools are predominantly step- (44.9%) and feather-terminated (33.8%) (Figure 106). Most (80.2%) are small in size. The ventral to dorsal scar ratio is .8:1, and the mean number of scars per tool is 131.5. The high number of scars is reflected in their distribution, which is one of contiguous, overlapping scars with smaller ones superposed over larger scars.

Edge Rounding: The tool edges are blunted, with slight to moderate localized rounding.

Polish: Dull to dull-bright, grainy polish is present on the edges and edge-aspects of all three tools. Tool 12-11 also has bright-grainy polish and traces of bright, smooth polish.

Polish (SEM): Smooth-pitted polish is mainly found on projections on the edge, ridges and scar boundaries of tool 12-11 (Figures 106,107).

Striations: One medium, narrow striation, oriented parallel to the edge, is present on polish on the ventral aspect of tool 12-11.

Striations (SEM): Short striations oriented parallel and oblique to the edge are visible at high magnifications on tool 12-11.

Comparisons: Odell (1980b) used several unretouched Cache Creek basalt tools to dismember a raccoon with slicing and sawing actions. The wear observed on the tools consisted primarily of bifacially distributed microflake scars, forming a denticulated edge. The scars were small to large in size and primarily hinge, step and feather terminated. Kamminga (1978:476,478) used altered and andesitic basalt tools to butcher a kangaroo and observed edge dulling and blunting as well as microflake scars. In the present experiments, small step and feather scars dominate the edge damage and some edge rounding is also present, with polish and striations observed at higher magnifications.

7.7.6 Defleshing Bone (Scraping)

Two tools made of vitreous basalt (Tools 44-8, 44-11; Figure 137e,f) were used to scrape remnants of flesh from fallow deer bones. The tools were used for 7 to 28 minutes (300-1000 strokes) with medium to heavy applied pressure to scrape scraps of flesh from the bones, heavily contacting the bone in the process. One of the tools (44-8), a concaved-edged tool, was perfectly suited for scraping the flesh off ribs, although it was somewhat in need of resharpening after 28 minutes of use (1000 strokes).

Microflake Scars: Step-terminated microflake scars (46.5%) are the most common type present on the tools, with lesser percentages of hinge (28.5%) and feather (24.3%) scars (Figure 108). Approximately three-quarters (74.1%) of the scars are small in size. On the dorsal edge-aspect scars tend to be deep, while they are predominantly shallow on the ventral edge-aspect. The ventral to dorsal microflake scar ratio is .4:1, and the mean number of scars per tool is 71.5. Scars are scattered to contiguous on the ventral edge-aspect and contiguous, overlapping and superposed on the dorsal.

Edge Rounding: Absent.

Polish: Polish is not common on either tool, being restricted to a few high points of the microtopography. On tool 44-11 it is dull, grainy-texured and present mainly on the dorsal edge-aspect with traces on the edge and ventral edge-aspect. On tool 44-8, the polish is present only on the edge and is dull to bright and grainy.

Polish (SEM): Polish is largely restricted to certain high points on the edge and dorsal edge-aspect of tool 44-8, although some is also present on parts of the ventral aspect. The polish has a smooth texture, with a flowing appearance, and lacks pits (Figure 107). Smoothed, but unpolished areas were also noted on both edge aspects. The largest heavily polished patch is 100 µm long and 80 µm wide. Smoothing extended .2 mm above the edge on the dorsal aspect, and a much lesser distance on the ventral aspect. Cracks are visible in the polish at 3500 x.

Striations: Absent.

Striations (SEM): Absent (44-8).

7.7.7 Defleshing Bone (Slicing)

Individual tools were used for 5 minutes to over 17 minutes (210 to 1000 strokes) on fallow deer ribs and a front foreleg. Three of the tools were made of non-vitreous basalt (Figure 137a-c) and one was vitreous basalt (Figure 137d) (Tools 2-1, 2-3, 11-3, 12-1). Flesh and fat were cut off the ribs with a slicing action using a medium to heavy pressure. A slicing action was also used to dissassemble the foreleg. Unavoidable heavy bone and cartilage contact was sustained by all tools, although flesh and fat were the primary contact materials. The vitreous basalt tool was dull but still usable after about 18 minutes of slicing (1000 strokes), but the non-vitreous ones were dulled after much less use.

Microflake Scars: Feather- (46.4%) and step-terminated (33.6%) microflake scars are the most common types (Figure 103). The vast majority (96.6%) of microflake scars are uniformly small. The ventral to dorsal flake scar ratio is .9:1 and the mean number of flakes per tool is 36.3. This low number is probably due to the difficulty encountered in recognizing scars on the predominantly coarse basalt used in these experiments. Microflake scars are predominantly scattered, either singly or in small groups of overlapping scars. On the most heavily used tool (vitreous basalt) the scars are contiguous, overlapping and often superposed (Figure 110).

Edge Rounding: Edges are unmodified to slightly blunted or rounded in places.

Polish: Extensive rounding and smoothing of the edges are visible at high magnifications on non-vitreous tools, but the surfaces of the smoothed areas are identical in brightness and texture to the unmodified surface of the basalt. The edge of the vitreous basalt tool (12-1) is not rounded, but traces of dull, grainy and bright, grainy polish are present on high points of the edge and both aspects. Slight smoothing of both edge aspects is also present.

Polish (SEM): Polish on tool 12-1 is scattered and restricted to a few high points and projections along the edge and extends slightly onto either aspect. It is smooth-textured and lacks true pits (Figure 111). In addition to the rare polished areas, there is extensive smoothing of the edge aspects. The polish is restricted in distribution to within 20 µm of the edge, while the smoothing extends 50 µm to 100 µm from the edge.

Striations: Absent.

Striations (SEM): Only one possible striation was observed on tool 12-1, and this at 4700 x. It is oriented oblique to the edge and measures approximately 2 um long.

7.8 Wood (N=29)

7.8.1 Scraping Dry Bark and Wood

All tools used to scrape dry bark and wood were made of vitreous basalt (Tools 27-11, 36-5, 44-10, 141-6, 141-11, 161-20; Figures 141d, 142a,d, 143c, 145a). The woods worked were Saskatoon, choke cherry, dogwood and balsam poplar which had been cut down between 10 days and 5 weeks prior to experimentation. Tools were used from 4 to 17 minutes (500-1150 strokes) to scrape wood (mainly with bark intact) on relatively thin diameter trees or branches, from which smaller branches were removed prior to experimentation. The activity is similar to the aboriginal debarking and smoothing of potential arrow or spear shafts. Medium to heavy pressure was applied to the tools. It is noteworthy that dry bark and wood are substantially harder to work than fresh wood, and that a considerable amount of heat was generated on the tool edges during the experiments.

Microflake Scars: Feather- (44.1%) and step-terminated (34.5%) microflake scars are the most common types present (Figure 112), although a substantial percentage (20.9%) of scars are hinge-terminated. Most scars (93.8%) are small in size. The ventral to dorsal microflake scar ratio is .2:1, and the mean number of scars per tool is 90.7. Scars on the ventral edge-aspect are rare or absent on individual tools, and when present are scattered, either singly or in small clusters of overlapping scars. Scars on the dorsal edge-aspect have a very high density and are contiguous and overlapping along the entire used edge. Also, small scars are superposed over larger ones.

Edge Rounding: The edges are slightly to moderately rounded (Figures 113,114).

Polish: Polish is present on the edges and edge-aspects of all tools, although a few of the lightly used ones only have traces. Bright, smooth, pitted polish is often continuous or nearly so along the edges and is mainly restricted to high points on the dorsal edge-aspects. The bright, smooth polish is often surrounded by dull to bright, grainy polish.

Polish (SEM): Polish is continuous along the edge of tool 141-6 and is present on both edge aspects, although it extends farther up the dorsal aspect. The polish is domed, smooth, and lacks true pits (Figures 106,107). It has a "smeared on" appearance, and at very high magnifications (2500 x and up) cracks become visible on the polish surface. The polish extends from 40 µm to 90 µm up the dorsal aspect, but is mainly in the 40 µm to 50 µm range.

Striations: Short, narrow, shallow striations are present on three tools and possibly a fourth. They are only found on polish and are oriented at or near right angles to the edge (70-120). They are very common on two tools.

Striations (SEM): Striations are relatively common on the polish of tool 141-6, but they are not visible at magnifications under 2500 x (Figure 114). They are narrow, of varying depths, and measure up to 14 μm long. They are mainly oriented at right angles to the edge, although some have an oblique orientation.

Comparisons: Schutt (1982:Appendix 2) used 12 glassy basalt tools to scrape dry poplar and dry pine. She observed polish on more than half of the tools and nearly all of them had edge rounding. Microflake scars were mainly present on the trailing edge-aspects and largely consisted of feather and step scars. This wear pattern is very similar to that observed on the tools in the present experiments, although Schutt did not observe striations.

7.8.2 Scraping Fresh Bark and Wood

Fresh Saskatoon and dogwood were scraped with 1 non-vitreous (Figure 143e) and 6 vitreous basalt tools (Figures 141c,f, 142b,c, 143d, 144a) (Tools 26-10, 27-15, 134-23, 138-6, 138-16, 141-4, 141-17). The trees had been cut down from 3 hours to 2 days prior to the experiments. Individual tools were used for approximately 2 to 69 minutes (200-5850 strokes). The tools were very effective for debarking and thinning-smoothing the wood (Figures 18,19), with little dulling on even the most heavily used tools. Medium to heavy pressure was applied to the tools.

Microflake Scars: The most common microflake scar type present on the tools is the step scar (57.4%), although feather-terminated scars are also common (35.7%) (Figures 115,117). The vast majority (92.8%) of microflake scars are small in size. The ventral to dorsal scar ratio is .08:1 and the mean number of scars per tool is 60.3. The rarely occurring scars on the ventral edge-aspect are widely scattered. Microflake scars are relatively dense on the dorsal edge-aspect and occasionally occur continuously, in overlapping and superposed configurations. More often the scars are found in small, scattered clusters, also in overlapping and superposed configurations.

Edge Rounding: Edges are slightly to moderately rounded (Figures 68,69, 116,118).

Polish: Bright, grainy polish with small, smooth, pitted-lined patches is virtually continuous along the edge of non-vitreous basalt tool 26-10. On the vitreous basalt tools, polish ranges from bright and grainy to very bright, and smooth-pitted (Figures 66-69). Large patches of smooth-pitted polish on the edge are usually surrounded by the grainy polish. Polish is mainly located on the rounded edges, but is often also present on both edge-aspects, usually to a lesser extent. Trend lines oriented at right angles to the edge are present.

Polish (SEM): Polish is rarely visible below 1000 x on tools 27-15 and 141-17. It is scattered in small patches and is primarily located on the very edge of the tool, and extends slightly more onto the ventral aspect than dorsal (Figures 116,118,119. The polish is very smooth in texture and has the appearance of having been "smeared on". At very high magnifications the polish surface has a "pebbly" appearance (Figure 121). The polish extends a maximum of 30 µm onto the ventral aspect and 20 µm onto the dorsal. Smoothing or light polish extends up to .1 mm from the edge on the ventral aspect.

Striations: Striations are present on two vitreous basalt tools (27-15, 134-23), and possibly a third (141-4). The striations are located on polish on the tool edges and are oriented at or near right angles to the edge. They are short, narrow and only abundant on tool 134-23 (Figures 67, 68).

Striations (SEM): Striations are uncommon on tools 27-15 and 141-17 and are only visible at very high magnifications on polished areas (Figure 119). They are oriented at or near right angles to the edge and are short and narrow. The longest striation measured 3 µm.

Comparisons: Montgomery (1978:273,274, Figure 68) used two vitreous basalt tools to scape a fresh Douglas fir branch. The microwear she observed consisted of small, uniform scars, mostly on the trailing edge aspect. Kamminga (1978:651) used an altered basalt tool to scrape fresh, dense wood and observed snap scars, edge rounding and smoothing. Schutt (1982:Appendix 2) used a few glassy basalt tools to scrape fresh poplar. She observed microscarring mainly consisting of feather and snap scars, as well as edge rounding and polish. In the present experiments most scars are step or feather terminated, with few snap scars. Overall, however, the wear observed by these researchers is not significantly different than the wear observed in the present study.

7.8.3 Sawing Fresh Wood

A debarked, 1 cm diameter Saskatoon branch cut down 6 hours prior to experimentation was sawn using vitreous basalt tools (Tools 27-10, 134-18; Figures 143a, 145f). They were used between 2 and 6 minutes (400-1000 strokes) with heavy applied pressure. The tools were used in a back and forth sawing motion at right angles to the long axis of the branch.

Microflake Scars: The most common microflake scar type is feather-terminated (47.7%), with step- (27.3%) and snap-terminated (19.3%) scars (27.3%) also being well represented. A large majority (84.1%) of the scars are small in size. The ventral to dorsal scar ratio is .8:1, and the mean number of scars per tool is 44. Microflake scars tend to be scattered, either singly or in small clusters with overlapping and occasionally superposed configurations.

Edge Rounding: Slight rounding is visible at high magnifications with the SEM (Figure 122).

Polish: On tool 27-10 there are scattered indications of dull to bright, grainy polish and bright, grainy to smooth polish on a few high points of the edge, with only slight traces on the edge-aspects. The more heavily used tool 134-18 has scattered bright, smooth-pitted polish surrounded by more extensive dull to bright, grainy polish along the edge with a nearly continuous distribution on both edge-aspects.

Polish (SEM): Polish is continuous along the edge of tool 134-18 and extends onto both aspects. It is first visible at about 250 x and has a smooth, but pitted texture (Figures 122,123). On the polish boundary it often has a "leafy", discontinuous distribution. As with wood polishes on other tools, this polish has a "smeared on" appearance. Smoothing or light polish is also present and extends up to 70 µm from the edge on both aspects. The highly developed polish extends a maximum of 50 µm from the edge.

Striations: Several short striations are present in the polish on the ventral edge-aspect of tool 134-18. These are oriented parallel or oblique to the edge.

Striations (SEM): No definite striations were observed on the tool 134-18, however, a few possible striations were noted at magnifications in excess of 2000 x.

Comparisons: Montgomery (1978:274, Figure 68) used one glassy basalt tool to saw fresh Douglas fir. She observed bifacially distributed microflake scars and a jagged edge. Kamminga (1978:590) used one olivine basalt tool to saw fresh, light-density wood, which resulted in a jagged and blunt edge, with some snap scars. Schutt (1982:Appendix 2) used a glassy basalt tool to saw green poplar. Edge rounding was present, as were bifacially distributed step, feather and snap scars. Microwear on the present experimental tools is very similar, with bifacially distributed feather, step and snap scars. Polish and striations are also present, but were not observed in the earlier studies, probably due to the low magnifications employed.

7.8.4 Sawing Dry Wood

Two vitreous basalt tools (Tools 63-2, 82-3; Figure 145 b,d) were used to saw 1 month old Saskatoon and 3 year old Pondorosa pine branches. Both branches were debarked shortly after they were cut down. The Saskatoon branch was 1 to 2 cm in diameter and the Pondorosa pine was 2 to 3 cm in diameter. Tools were used for approximately 3 to 4 minutes with heavy applied pressure to saw through the branches.

Microflake Scars: The most common microflake scar termination types are feather (44.1%) and step (43.2%) scars. Over three-quarters (76.8%) of the scars are small in size. The ventral to dorsal microflake scar ratio is .7:1 and the mean number of scars per tool is 56. Microflake scars are scattered singly or in small clusters in overlapping and occasionally superposed configurations.

Edge Rounding: Tool edges are slightly rounded or blunted in places.

Polish: Only slight traces of polish are present on the edge of tool 63-2. There are scattered occurrences of bright, smooth polish on high points of the edge of tool 82-3 with traces of polish on both edge-aspects.

Striations: Absent.

Comparisons: Schutt (1982:Appendix 2) used several glassy basalt tools to saw dry pine and observed polish and edge rounding on all of them. She also found bifacially distributed microscars which were mainly feather, step or snap terminated. This wear pattern is very similar to that observed for dry wood sawing in the present experiments.

7.8.5 *Sawing Bark*

Birch bark was removed from a tree 2 weeks prior to the experiment, while the tree itself had been cut down another 2 weeks earlier. A flat to slightly curved bark sheet was cut into strips with a sawing motion, starting at the top edge of the sheet. Medium pressure was applied to the vitreous basalt tool (Tool 27-7; Figure 143b) used in this experiment. The tool was used for 6 minutes (600 strokes). The bark was successfully cut into strips.

Microflake Scars: Feather-terminated scars are the most common microflake scar type (44.0%), but both snap (32.0%) and step scars (24.0%) are well-represented. The majority of scars (82.0%) are small. Prominent on the edge are medium to very large snap terminated scars, in which many of the other scars area found. Use-retouch is alternating. The ventral to dorsal scar ratio is .9:1 and there are 50 microflake scars on the tool. Microflake scars occur singly or in small clusters in an overlapping configuration.

Edge Rounding: Absent.

Polish: Bright, smooth, unpitted polish is mainly found on high points of the edge and edge-aspects of this tool. Bright, grainy and dull, grainy polishes are also present on the edge, often adjacent to the bright, smooth polish.

Striations: Absent.

7.8.6 *Slicing Bark*

One vitreous basalt tool (Tool 27-14; Figure 144f) was used to slice birch bark on a birch tree which had been felled approximately 2 weeks before the experiment began. The purpose was to cut the bark so that a sheet could be removed from the tree (this sheet was used in bark sawing experiment 7.8.5). The tool edge was held nearly vertical to the contact surface and was used with medium to heavy applied pressure. The tool was effective at the task, which took 6 minutes (150 strokes) to accomplish.

Microflake Scars: Feather- (37.4%) and hinge-terminated (36.3%) scars are the most common scar types. Most scars (85.7%) are small in size. The ventral to dorsal flake scar ratio is .9:1 and there are 91 flake scars on the tool. They are mainly contiguous and overlapping.

Edge Rounding: The edge is slightly rounded.

Polish: At 300 x the edge appears very rounded to facetted, with nearly continuous dull, grainy to bright, grainy polish. Polish on the dorsal and ventral edge-aspects is scattered but plentiful, usually located near the edge, but occasionally penetrating some distance from it. This polish is dull and grainy and mainly located on high points.

Striations: Absent,

7.8.7 *Whittling Fresh Wood and Bark*

Saskatoon branches cut down 2 days prior to experimentation and willow branches cut down minutes before experimentation were used in these experiments. Bark was left on the branches but small twigs were removed prior to experimentation. The bark and some wood was whittled off the willow branches, with debarking being the goal of the experiment. The Saskatoon branches were whittled to points. One non-vitreous (Figure 140f) and five vitreous basalt tools (Figures 141a,b,e 144b, 145e) were used between 2 and 12 minutes (approximately 200-500 strokes) in the experiments (Tools 12-5, 17-12, 80-9, 80-10, 80-11, 23-4). Medium to heavy pressure was applied to the tools, with one being partially wrapped in rawhide to protect the experimenter's hand. The tools were very effective in accomplishing the tasks and still usable after even 12 minutes of use.

Microflake Scars: Just over half (52.0%) of the microflake scars are feather-terminated, although a substantial percentage (32.5%) are step-terminated (Figure 124). A substantial majority (85.2%) of scars are small in size. The ventral to dorsal microflake scar ratio is .3:1 and the mean number of scars per tool is 42. Scars are rare and mainly scattered singly on the ventral edge-aspects, although clusters of a few overlapping scars are also present. On the dorsal edge-aspects scars are much more common and are more or less continuous along the used edge with overlapping and superposed configurations.

Edge Rounding: Edges are blunted and occasionally rounded (Figure 125).

Polish: Polish is present on all of the vitreous basalt tools. Bright, smooth-pitted polish is present on the edges, and varies in distribution from nearly continuous to scattered traces on different tools. Dull, grainy and bright, grainy polishes are also present, as is bright, smooth, unpitted polish. Polish is always more extensive on the dorsal edge-aspect than ventral, and has the same characteristics as that found on the edge. It is not possible to distinguish polish on the non-vitreous tool 23-4.

Polish (SEM): Polish on the edge of tool 12-5 is continuous, bifacial and extensive, but more pronounced on the dorsal edge-aspect than the ventral (Figure 125). It is very smooth-textured and has the typical "smeared on" appearance of polish on wood working tools. The highly developed polish usually extends 20 µm to 30 µm from the edge. A zone of lighter polish or smoothing usually extends from 40 µm to 60 µm from the edge.

Striations: Possible striations are present in the polish on the edge of two tools (80-9, 80-10). They are oriented between 90 and 115 degrees to the edge.

Striations (SEM): A few possible striations were noted on tool 12-5 at very high magnifications.

Comparisons: Schutt (1982: Appendix 2) used one glassy basalt tool to whittle green poplar. She observed edge rounding and bifacial microflake scars (feather, step and snap), but no polish or striations. With the exception of the polish and striations, this wear pattern is similar to that observed in the present study.

7.8.8 Whittling Dry Wood

Balsam poplar and Saskatoon cut down one month prior to experimentation were used in the dry wood whittling experiments. The balsam poplar branch was first debarked by whittling in one experiment and was whittled to a point in another. One experiment involved thinning the end of Saskatoon branches. Vitreous basalt tools (Tools 27-8,44-7, 44-9, 161-15; Figures 143f, 144c-e) were used for 4 to 13 minutes (200-550 strokes). The tools were found to be effective whittling implements.

Microflake Scars: There are nearly equal percentages of feather- (38.7%) and step-terminated (35.8%) microflake scars. Less common in general are hinge-terminated scars (19.0%), which, however, are very common (39.0%) on the tool used to whittle Saskatoon. Three quarters (75.6%) of the scars are small in size. The ventral to dorsal microflake scar ratio is .7:1 and there is an average of 110.8 scars per tool. On the ventral edge-aspect microflake scars are scattered in small groups of scars in overlapping and superposed configurations. On the dorsal aspect, scars are continuous or nearly continuous along the used edge, also in overlapping and superposed configurations.

Edge Rounding: Edges are rounded or blunted.

Polish: Scattered patches of bright, smooth-pitted polish are associated with usually more extensive dull to bright, grainy areas along the edges and both edge-aspects of the tools. Bright, smooth, but unpitted polish is also present.

Striations: On tool 27-8 numerous short striations are present on the polish and are oriented at or near right angles to the edge.

Comparisons: Schutt (1982:Appendix 2) used one glassy basalt tool to whittle dry pine. She didn't observe polish, edge rounding or striations. As in the present experiments microflake scars were bifacially distributed, although primarily on the ventral aspect, and were mainly feather- or step-terminated.

CHAPTER 8

STATISTICAL ANALYSIS OF MICROFLAKE SCAR VARIABLES

8.1 *Introduction*

Statistical techniques have only occasionally been applied to microwear research problems (see Nance 1977; Stafford 1977; Odell 1977; Siegel 1984; Vaughan 1985; Schutt 1982; Lewenstein 1981), although much of the observational data is amenable to such analyses. In this chapter the utility of two multivariate techniques for objective functional classification is evaluated. In addition, the usefulness of microflake scar attributes is assessed.

All statistical analyses were undertaken at the University of Saskatchewan with SPSS-X (Norusis 1985) on a VAX 8600 computer using version 4.3 of the VMS operating system.

8.2 *Correlations Between Independent and Dependent Variables*

8.2.1 *Introduction*

The Spearman Rank Order correlation coefficient (Leach 1979:190-194; Thomas 1976:395-405) was used to determine the strength and direction of relationships between selected independent and dependent variables. A non-parametric correlation coefficient was chosen because the underlying assumptions of parametric measures can not be met with the data. The variables are defined on Tables 10 and 11 and the Spearman correlation coefficients are presented on Tables 12 and 13.

Table 10. Dependent variable abbreviations and definitions.

Abbreviation	Definition
1. DSTEPS	Number of small dorsal step scars.
2. DSTEPM	Number of medium dorsal step scars.
3. DSTEPL	Number of large dorsal step scars.
4. DSTEPV	Number of very large dorsal step scars.
5. DHINGES	Number of small dorsal hinge scars.
6. DHINGEM.	Number of medium dorsal hinge scars.
7. DHINGEL	Number of large dorsal hinge scars.
8. DHINGEV	Number of very large dorsal hinge scars.
9. DFEATHS	Number of small dorsal feather scars.
10. DFEATHM	Number of medium dorsal feather scars.
11. DFEATHL	Number of large dorsal feather scars.
12. DFEATHV	Number of very large dorsal feather scars.
13. DSNAPS	Number of small dorsal snap scars.
14. DSNAPM	Number of medium dorsal snap scars.
15. DSNAPL	Number of large dorsal snap scars.
16. DSNAPV	Number of very large dorsal snap scars.
17. VSTEPS	Number of small ventral step scars.
18. VSTEPM	Number of medium ventral step scars.
19. VSTEPL	Number of large ventral step scars.
20. VSTEPV	Number of very large ventral step scars.
21. VHINGES	Number of small ventral hinge scars.
22. VHINGEM	Number of medium ventral hinge scars.
23. VHINGEL	Number of large ventral hinge scars.
24. VHINGEV	Number of very large ventral hinge scars.
25. VFEATHS	Number of small ventral feather scars.
26. VFEATHM	Number of medium ventral feather scars.
27. VFEATHL	Number of large ventral feather scars.
28. VFEATHV	Number of very large ventral feather scars.
29. VSNAPS	Number of small ventral snap scars.
30. VSNAPM	Number of medium ventral snap scars.
31. VSNAPL	Number of large ventral snap scars.
32. VSNAPV	Number of very large ventral snap scars.
33. VTOTSMAL	Total number of small ventral scars.
34. VTOTMED	Total number of medium ventral scars.
35. VTOTLARG	Total number of large ventral scars.
36. VTOTVLRG	Total number of very large ventral scars.
37. DTOTSMAL	Total number of small dorsal scars.
38. DTOTMED	Total number of medium dorsal scars.
39. DTOTLARG	Total number of large dorsal scars.

continued

Table 10. (continued)

Abbreviation	Definition
40. DTOTVLRG	Total number of very large dorsal scars.
41. VTOTSTEP	Total number of ventral step scars.
42. VTOTHING	Total number of ventral hinge scars.
43. VTOTFEAT	Total number of ventral feather scars.
44. VTOTSNAP	Total number of ventral snap scars.
45. DTOTSTEP	Total number of dorsal step scars.
46. DTOTHING	Total number of dorsal hinge scars.
47. DTOTFEAT	Total number of dorsal feather scars.
48. DTOTSNAP	Total number of dorsal snap scars.
49. SMALLPER	Percentage of small scars on tool regardless of type.
50. MEDPER	Percentage of medium scars on tool regardless of type.
51. LARGEPER	Percentage of large scars on tool regardless of type.
52. VLARGPER	Percentage of very large scars on tool regardless of type.
53. STEPPER	Percentage of step scars on tool regardless of size.
54. HINGEPER	Percentage of hinge scars on tool regardless of size.
55. FEATHPER	Percentage of feather scars on tool regardless of size.
56. SNAPPER	Percentage of snap scars on tool regardless of size.
57. HINGSTEP	Percentage of step and hinge scars combined on tool regardless of size.
58. DORSPER	Percentage of total number of scars of all types and sizes on dorsal aspect.

Table 11. Independent variable abbreviations and definitions.

SPINEANG:	Spine-plane angle of used part of tool. Rounded to nearest degree.
APPRES:	Pressure applied to tool during use. Pressure is ranked as follows: 1=Light 2=Light-to-Moderate 3=Moderate 4=Moderate-to-heavy 5=heavy
HARDTHRE:	Relative contact material hardness ranked on a 3-part scale: 1=soft (butchering, plants) 2=medium (hide, wood) 3=hard (bone, antler)
HARDFIVE:	Relative contact material hardness ranked on a 5-part scale: 1=soft (flesh) 2=medium-soft (fresh hide, plants) 3=medium-hard (fresh wood, dry hide) 4=hard (dry wood, soaked antler) 5=very hard (unsoaked bone and antler)
SPECONT:	Specific contact materials ranked in order of estimated hardness: 1=deer flesh and fat 2=fish flesh, bone and skin 3=fish flesh, bone, skin and scales 4=fish scales, skin 5=deer flesh, fat and hide 6=deer flesh, fat, hide and bone 7=green plant stems 8=dry plant stems 9=flesh side of hide 10=hair side of hide 11=deer hide, bone, cartilage, flesh and fat 12=deer bone and flesh 13=dry defleshed, dehaired hide 14=bark (birch) 15=fresh wood and bark 16=dry wood and bark 17=soaked antler 18=soaked bone 19=unsoaked antler 20=unsoaked bone

Only some independent variables lend themselves to ranking and the following were used in this study: spine-plane angle, applied pressure, contact material hardness ranked on a three-part relative hardness scale, contact material ranked on a five-part relative hardness scale, and all contact materials ranked in order of relative hardness (Table 11). All of the dependent variables chosen were attributes of microflake scars: scar size, scar termination type and scar location (dorsal or ventral aspect) (Table 10). Microflake scar attributes were chosen rather than other types of wear for several reasons: (1) the attributes could be objectively observed (unlike polish texture or reflectivity or amount of edge rounding); (2) the attributes were present on almost all of the tools (unlike striations, edge rounding or polish); and (3) ease of observation--the attributes can be observed with low magnifications (unlike polish or striations).

8.2.2 *Results of the Spearman Correlations*

The Spearman correlations only indicated significant relationships in three areas. These are discussed below.

There is a relationship between spine-plane angle and microflake scar termination type. The independent variable SPINEANG is correlated with the dependent variables VTOTSNAP, DTOTSNAP, DTOTSTEP, STEPPER and SNAPPER. There is a negative correlation between spine-plane angle and snap scars--tools with high spine-plane angles have fewer snap scars than tools with low spine plane angles (Tables 12, 13). There is also a positive correlation between spine-plane angle and step scars--tools with high spine-plane angles have more step scars than tools with low spine-plane angles (Tables 12, 13). These correlations support similar observations made by other researchers without the benefit of statistical analysis (e.g. Cotterell et al. 1979:134; Tringham et al. 1974:191). The explanation for these correlations is beyond the scope of this study, although high spine-plane angled edges have thicker, stronger edges than lower angled tools and would therefore be less prone to snapping during use. Lawrence (1979:119) noted that fractures on thick edges tend to have hinge or step terminations.

There is a relationship between contact material hardness and microflake scar size. The independent variables APPRES, HARDTHRE, HARDFIVE, SPECONT, which represent contact material hardness, are correlated with the dependent variables VTOTSMAL, DTOTSMAL, VTOTMED, VTOTLARG, VTOTVLRG, DTOTSMAL, DTOTMED, DTOTLARG, DTOTVLRG, SMALLPER, MEDPER, LARGEPER, and VLARGPER. There is a weak negative correlation between contact material hardness and small-sized microflake scars--tools used on harder contact materials tend to have fewer small scars than tools used on softer contact materials (Tables 12, 13). The correlation coefficient between small scar frequency and contact material hardness is only very weakly negative and in a few cases weakly positive. The correlation coefficient between percentage of small scars and contact material

Table 12. Spearman Rank Order Correlations between selected dependent and independent variables on all tools (N=90). Variable definitions can be found on Tables 10 and 11.

INDEPENDENT VARIABLE	DEPENDENT VARIABLE	CORRELATION COEFFICIENT
SPINEANG	VTOTSNAP	-.4867
	DTOTSNAP	-.5244
	DTOTSTEP	+.5043
APPRES	VTOTSMAL	+.0927
	VTOTMED	+.5381
	VTOTLARG	+.5848
	VTOTVLRG	+.3918
	DTOTSMAL	-.0769
	DTOTMED	+.3361
	DTOTLARG	+.3961
	DTOTVLRG	+.4003
HARDTHRE	VTOTSMAL	-.1291
	VTOTMED	+.2564
	VTOTLARG	+.4222
	VTOTVLRG	+.3952
	DTOTSMAL	+.1491
	DTOTMED	+.3887
	DTOTLARG	+.3745
	DTOTVLRG	+.4115
HARDFIVE	VTOTSMAL	-.0614
	VTOTMED	+.3192
	VTOTLARG	+.4083
	VTOTVLRG	+.4329
	DTOTSMAL	+.2083
	DTOTMED	+.3521
	DTOTLARG	+.3065
	DTOTVLRG	+.3793
SPECONT	VTOTSMAL	-.0550
	VTOTMED	+.3545
	VTOTLARG	+.4421
	VTOTVLRG	+.4183
	DTOTSMAL	+.2292
	DTOTMED	+.4187
	DTOTLARG	+.3509
	DTOTVLRG	+.4261
	DTOTHING	+.3311

Table 13. Spearman Rank Order Correlations for all tools (N=90). The variable definitions are presented on Table 10 and 11. Percentages of microflake scar type and size are used rather than raw counts as on Table 12.

INDEPENDENT VARIABLE	DEPENDENT VARIABLE	CORRELATION COEFFICIENT
SPINEANG	STEPPER	+.3860
	SNAPPER	-.6056
APPRES	SMALLPER	-.5295
	MEDPER	+.5082
	LARGEPER	+.5414
	VLARGPER	+.3780
HARDTHRE	STEPPER	+.1692
	HINGEPER	+.2792
	FEATHPER	-.3158
	SNAPPER	-.0350
	HINGSTEP	+.3393
	SMALLPER	-.4429
	MEDPER	+.3645
	LARGEPER	+.5151
	VLARGPER	+.4755
HARDFIVE	STEPPER	+.0911
	HINGEPER	+.2622
	FEATHPER	-.1797
	SNAPPER	-.0493
	SMALLPER	-.3977
	MEDPER	+.3751
	LARGEPER	+.4692
	VLARGPER	+.4433
SPECONT	STEPPER	+.1485
	HINGEPER	+.3029
	FEATHPER	-.2536
	SNAPPER	-.0311
	HINGSTEP	+.3355
	SMALLPER	-.4408
	MEDPER	+.4207
	LARGEPER	+.5137
	VLARGPER	+.4729

hardness is also not strongly negative (Tables 12, 13). There is also a weak positive correlation between contact material hardness and medium to very large sized microflake scars--tools used on harder contact materials have more large scars than tools used on softer contact materials (Table 12, 13). In summary, a tool used with heavy applied pressure on harder contact materials will result in more large scars being removed than a tool used with light pressure on softer contact materials. These results support a statistical analysis by Lewenstein (1981:182) and the non-statistical observations of other researchers (e.g. Odell and Odell-Vereecken 1980:110; Cotterell and Kamminga 1979:111; Tringham et al. 1974:189).

There is a relationship between contact material hardness and microflake scar termination type. The independent variables APPRES, HARDTHRE, HARDFIVE, and SPECONT are correlated with the dependent variables DTOTHING, FEATHPER, HINGEPER. There is a weak negative correlation between contact material hardness and feather-terminated microflake scars--tools used on harder contact materials tend to have a lower percentage of feather scars than tools used on softer materials (Table 13). There is a very weak positive correlation between contact material hardness and hinge-terminated microflake scars--tools used on harder contact materials have more hinge-terminated scars than tools used on softer contact materials (Tables 12, 13). These results again support non-statistical observations made by other researchers (e.g. Odell 1981:200,204; Odell and Odell-Vereecken 1980:101; Cotterell et al. 1979:133; Cotterell and Kamminga 1979:111; Lawrence 1979:119).

8.2.3 *Summary and Discussion of Correlation Results*

It should be noted that none of the correlations are particularly strong and thus only indicate weak relationships between some microflake scar attributes and independent variables. The Spearman correlations generally agree with observations made by other researchers regarding relationships between tool use and microflake scar attributes. However, one assertion made by earlier researchers is not supported: step scars are not correlated with contact material hardness, contrary to the opinions of Odell and Odell-Vereecken (1980:101), Tringham et al. (1974:189) and Lawrence (1979:119).

8.3 *Cluster Analysis*

8.3.1 *Introduction*

Cluster analysis (Sneath and Sokal 1973; Romesburg 1984) is commonly used in archaeology, but has only rarely been applied to microwear research. Nance (1977) used cluster analysis in an attempt to determine if groups of variables were correlated with morphological tool types in an assemblage of Palaeo-Indian

Table 14. Agglomeration methods and distance measures used for cluster analyses (see Norusis 1985:184).

Agglomeration Method	Distance Measure
Average Linkage Between Groups	Euclidean
	Squared Euclidean
	City Block
	Cosine
Average Linkage Within Groups	Euclidean
	Squared Euclidean
	City Block
	Cosine
Single Linkage	Euclidean
	Squared Euclidean
	Cosine
Complete Linkage	Euclidean
	Squared Euclidean
	Cosine

tools, while Siegel (1984) employed cluster analysis on an ethnographic collection of Eskimo endscrapers to determine the materials each was used on.

The problem with applying cluster analysis to an assemblage of tools of unknown function is that the meanings of the derived clusters are open to several interpretations (see Siegel 1984, 1986; Bamforth 1986; Hayden 1986). The uncertainty of the significance of the clusters, or doubt if meaningful clusters have even been produced, forces the researcher to re-introduce a subjective element into the study after using an objective method to group the tools.

Hayden (1986:67) recommends that multivariate statistical techniques be thoroughly tested on experimental tools to evaluate the usefulness of the algorithms and the variables before these techniques are applied to the analysis of archaeological materials. The cluster analyses discussed below and the discriminant function analyses discussed in Section 8.4 were undertaken with experimental data to evaluate the appropriateness of these statistical techniques for functional classification and to identify functionally significant variables.

8.3.2 *Cluster Analysis Results*

Microflake scar variables on 90 vitreous basalt tools were subjected to several combinations of clustering methods and distance measures (Table 14). Several combinations were more or less successful in separating the tools into groups having functional meaning, although the single linkage (or nearest neighbour) method produced "chaining" (Doran and Hodson 1975:176; Sneath and Sokal 1973:223) and was unusable. The complete linkage (or furthest neighbor) method consistently produced the best results regardless of the distance measure used. Cosine consistently proved to be the best distance measure of several available with SPSS-X (see Norusis 1985:184), and probably worked best because of its propensity for ignoring size displacements in variables, thus grouping cases on the patterning in their variable profiles (Romesburg 1984:100-101). Converting the variables into standardized Z-scores made very little difference to the results.

Figure 29 is the dendrogram resulting from clustering the 32 raw flake variables using cosine as the distance measure and complete linkage as the agglomeration method. This dendrogram is comprised of 5 major clusters (clusters A-E) which are discussed in terms of both the independent variables (e.g. action, contact materials, number of use-strokes) and dependent variables (e.g. microflake scar attributes) of the tools. The independent variables are discussed first.

It can be seen that major clusters A and C are almost exclusively comprised of tools used with transverse actions (31 out of 38 or 81.6%) while major clusters D and E are almost exclusively comprised of tools used with longitudinal actions (39 out of 47 or 83.0%) (Figure 29). Cluster B consists of a small (n=5), mixed group of longitudinally (60%) and transversely (40%) used tools. The tools in

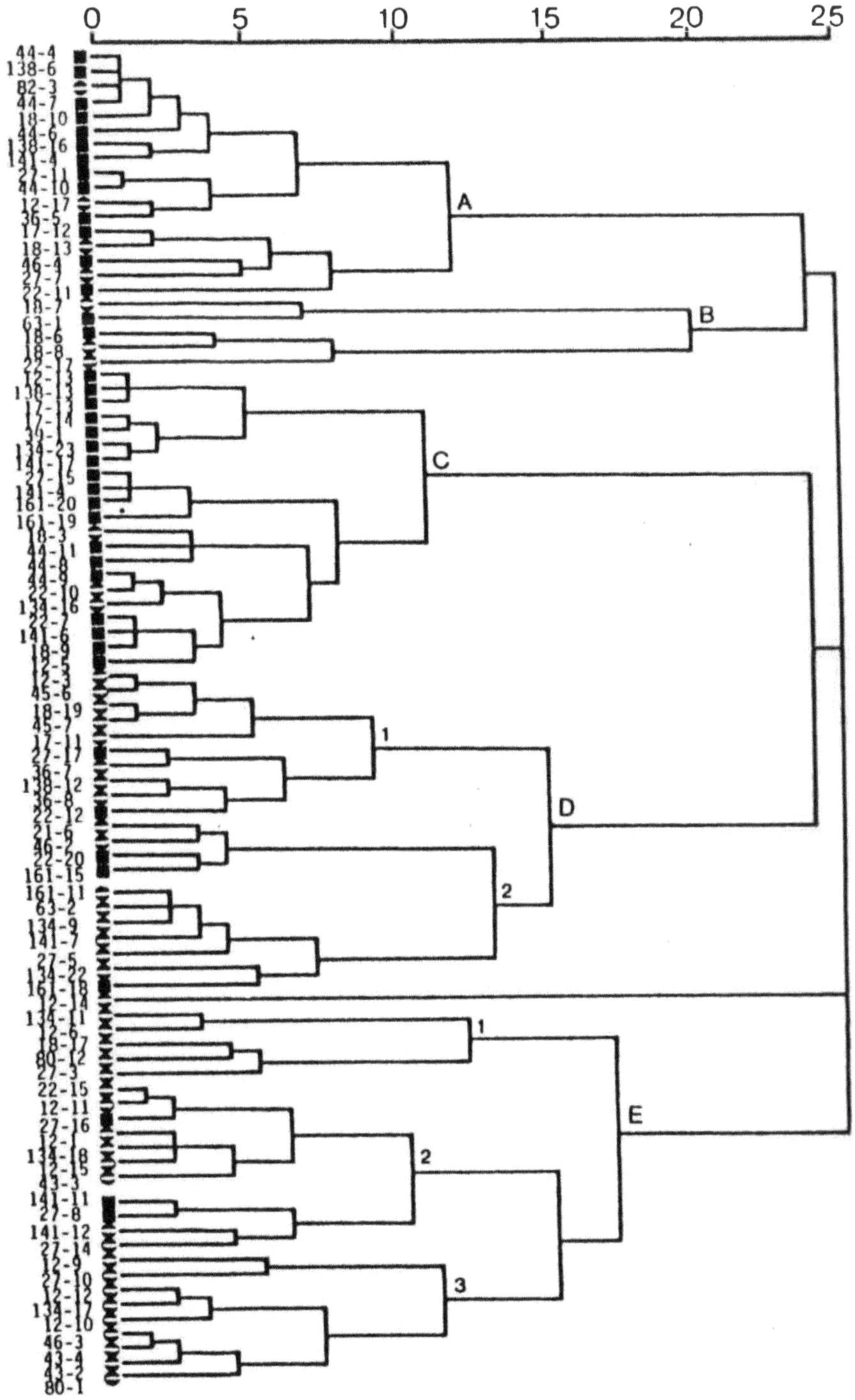

Figure 29. Dendrogram produced by clustering 90 vitreous basalt tools using 32 microflake scar variables (see Table 10:1-32). Cosine was the distance measure employed and complete linkage was the agglomeration method used. Each tool is identified by its number on the far left. Tools used with transverse actions are indicated by black rectangles and tools used with longitudinal actions are indicated by split circles.

cluster B are all very lightly used (60 to 100 strokes) and therefore probably have not developed scar patterns clearly reflecting their use, preventing them from being grouped with tools used with the same actions.

The clusters have some meaning with regard to contact material ranked on the basis of hardness. Cluster A is comprised almost entirely of tools used transversely on medium contact materials (wood and hide), while Cluster C is mainly comprised of tools used on both medium (wood and hide) and hard contact materials (bone and antler). Cluster D1 is comprised of tools used longitudinally on soft and hard contact materials (flesh, bone and antler), while Cluster D2 is mainly comprised of tools used longitudinally on medium and hard contact materials (wood, hide, bone and antler). Clusters E1 and E2 are both comprised of tools used longitudinally on soft, medium and hard contact materials. Cluster E3 is mainly comprised of tools used longitudinally on soft contact materials (plants, flesh), although a few used on medium materials (hide, wood) are also present.

The ventral to dorsal microflake scar ratio (v:d ratio) (see Chapter 6.6.2) was not one of the dependent variables used in this analysis, however, microflake scar location (dorsal or ventral edge-aspect) was used and reflects the same information about scar distribution. Clusters A and C had v:d ratios typical of transversely used tools (Table 27) and clusters D and E had ratios characteristic of longitudinally used tools (Table 15).

Cluster C is comprised of tools used on medium and hard contact materials while cluster A tools were used primarily on medium contact materials. Microflake variables which reflect contact material hardness, such as termination type and size (see section 8.2) are distributed as follows between the two clusters: cluster A has a much higher percentage of feather-terminated scars, a lower percentage of hinge-terminated scars, and a higher percentage of small scars than cluster C (Table 15). Cluster D1 is comprised of tools used on both soft and hard materials, while cluster D2 consists of tools used on medium and hard contact materials. Therefore, cluster D1 dependent variables as a group should reflect an overall softer contact material aggregate than cluster D2. This is, in fact, the case: cluster D1 has a higher percentage of feather-terminated scars, a lower percentage of hinge-terminated scars and a higher percentage of small scars than cluster D2 (Table 15). Clusters E1 and E2 are comprised of tools used on contact materials of all hardnesses, while cluster E3 tools were mainly used on soft materials. This is reflected in the contact material sensitive variables: cluster E3 has a higher percentage of feather-terminated scars, a lower percentage of hinge-terminated scars and a higher percentage of small scars than clusters E1 and E2 (Table 15).

8.3.3 *Summary and Discussion of Cluster Analysis Results*

In summary, tools used with transverse or longitudinal actions were consistently grouped together. As well, tools used on similar contact materials tended to be grouped together within the major clusters of longitudinally and

Table 15. Selected dependent variable distribution summary for clusters A-E on the cluster analysis dendrogram (Figure 29).

CLUSTER	MICROFLAKE SCAR VARIABLES			
	% OF SMALL SCARS	% OF HINGE SCARS	% OF FEATH SCARS	V:D SCAR RATIO
A	87.5	8.7	60.8	.3
B	94.0	36.5	18.5	.6
C	83.6	9.1	24.1	.2
D1	88.2	7.8	35.8	2.4
D2	77.2	19.5	23.9	1.3
E1	83.5	10.7	32.4	1.1
E2	87.6	18.0	45.6	1.1
E3	88.5	2.7	58.0	1.9

Table 16. Dependent variable sets used for discriminant function analysis.

VARIABLE SET	DEPENDENT VARIABLES ON TABLE 10
8	41-48
16	33-48
32	1-32

transversely used tools. This analysis has demonstrated that the three microflake attributes of size, termination type, and location combined into 32 dependent variables *do* reflect the function of the tools and that it is possible to use objective, repeatable methods to determine tool function. These variables reflect tool action well and contact material only fairly.

As discussed in section 8.3.1, the advantage of employing cluster analysis to this assemblage is that since the functions of the tools are known, the usefulness of the clustering methods employed and the appropriateness of the selected variables for segregating the tools into functional types can be assessed. In this case, complete linkage proved to be the best agglomeration method, a result which is not unexpected (see Doran and Hodson 1975:176). Several distance measures resulted in functionally meaningfull clusters, but cosine produced the most clear-cut functional groupings. The clusters have the most functional meaning in regard to tool action and only secondarily in regard to contact material. Without prior knowledge it would be almost impossible to accurately infer contact material from any of the dendrograms produced.

It became apparent during this exercise that a more appropriate statistical method must exist for the objective functional classification of stone tools and for assessing the validity of using the selected microflake scar variables to identify various tool functions. A search of the literature indicated that discriminant function analysis would be potentially more suitable for addressing these problems than cluster analysis.

8.4 *Discriminant Function Analysis*

8.4.1 *Introduction*

The partial success of cluster analysis indicated the potential of multivariate statistical methods for objective functional classification. Discriminant function analysis (DFA) is a more sophisticated multivariate method which has been successfully applied in recent years to address several archaeological problems requiring the objective classification of cases (e.g. Chandler and Ware 1976; Burton 1980; Walker 1980; Greaves 1982; Magne 1985; Walde 1985). DFA has only been used in one previous experimental microwear research project (Stafford 1977) and its application here is an attempt to determine the suitability and potential of the method to provide objective functional classifications of stone tools based on selected microwear variables. At the same time, DFA will provide an evaluation of the usefulness of the chosen variables (Tables 10,11,16) as indicators of tool action and contact material hardness. Stepwise selection of variables was employed using the selection criterion of minimizing Wilk's lambda (Norusis 1985:73-122).

Discriminant function analysis is used to classify cases (in this case stone tools) into mutually exclusive groups on the basis of the characteristics of these cases (Norusis 1985:73; Afifi and Clark 1984:247). This method differs from cluster analysis, factor analysis and multi-dimensional scaling, which are "unsupervised learning technique" used to search for patterns or groupings among cases (Wolff and Parsons 1983). DFA is a "supervised learning technique" (Wolff and Parsons 1983) which uses cases of known group membership to determine linear combinations of variables which result in the best separations between the groups (Norusis 1985:73). Once this has been accomplished, cases of unknown group membership can be classified with a known degree of accuracy. Further, the variables which contribute most to the classification can be identified (Norusis 1985:73; Afifi and Clark 1984:247), which allows the dropping of unimportant variables (see Magne 1985 for an archaeological example).

There are certain requirements of the data which are recommended to be met if the linear discriminant function is to be optimal, minimizing the chances of misclassification (Norusis 1985:75). Archaeological data often do not meet these requirements; fortunately, the method is robust enough to still be effective, as long as the data are not in extreme violation of the requirements (Hair et al. 1979:86-87).

The accuracy of the discriminant functions to segregate cases into groups in the present study was estimated using the following two methods:

1. Empirical Method (Afifi and Clark 1984:266): This is the standard means of assessing the validity of the discriminant function (DF). The DF is applied to the same sample of cases from which it was derived and the results are presented in tabular form, showing the percentage of cases correctly classified. This estimate is biased and often overestimates the actual degree of accuracy because the same sample is used for deriving and validating the DF.

2. Cross-Validation (Afifi and Clark 1984:266): To circumvent the bias problem inherent in the "empirical" method and arrive at an unbiased estimate, the DF is derived from one sample and applied to another.

In this study the accuracy of the DFs were first estimated using the empirical method and then "validated" by classifying 10 blind test tools (see Chapter 9) not used to derive the DFs, resulting in what are considered to be reliable, unbiased accuracy estimates.

The importance of individual variables in deriving a discriminant function can perhaps be best assessed by examining the correlation (Pearson correlation coefficient) between the DF and the dependent variables (Hair et al. 1979:104-105; Norusis 1985:91-92). The dependent variables with the highest correlations with each DF are listed on Table 18, and their significance is discussed in following sections. Three "sets" of dependent variables were employed in the DFA "runs" (Table 16).

8.4.2 *Tool Action*

The accuracy of DFA using three different variable sets can be seen on Table 17. For discriminating longitudinally vs transversly used tools it makes very little difference if the 8, 16 or 32 variable sets are employed (Tables 10, 16). Of interest is the fact that the 8-variable set is as accurate as the others although each variable has measurements on only 2 dimensions, microflake scar termination type and location, ignoring the third dimension, scar size. Feather-termination, small size and ventral edge-aspect location are the variables most highly correlated with the DF (Table 18).

There is no difference in accuracy between the 16 and 32 variable sets for distinguishing cutting, whittling and scraping actions, and they are both significantly better than the results from using the 8 variable set (Table 17). Apparently scar size, which is not represented in the 8-variable set, contributes to distinguishing between the three actions. Feather- termination, small size and very large size are the scar variables most highly correlated with the DFs (Table 18).

To summarize, feather-termination and small size are the two microflake scar variables most important for discriminating between tool actions. There is a slight preference for ventral location, especially for discriminating between longitudinal and transversely used tools. A high degree of accuracy in discriminating between two, or even three actions, solely on the basis of microflake scar characteristics, has been demonstrated. These results were confirmed by classifying blind test tools of unknown function with the same DF (Table 19). In fact, the blind test results were slightly better than the experimental tool results.

8.4.3 *Contact Material*

Only the 16 and 32 variable sets (Table 16) were used in the attempt to discriminate between tools on the basis of contact material. Two measures of contact material hardness were employed: three- and five-part relative hardness scales (Table 11) based on those of Tringham and Odell (Tringham et al. 1974; Odell and Odell-Vereecken 1980).

The results for the 32 and 16 variable sets for discriminating contact materials on the 3-part scale were quite good, although significantly higher for the 32 variable set (Table 17). The 16 variable set was, therefore, not used in the rest of the analyses. Step- and feather- termination, small size and medium size are the variables most highly correlated with the DFs (Table 18). There is no preference for scar location.

Table 17. The percentage of tools corrrectly classified using three variable sets (see Tables 10 and 16).

VAR		% CORRECTLY CLASSIFIED		
SET	LorT	C-W-S	H-3	H-5
32	86.67	80.00	81.11	70.00
16	87.78	80.00	72.22	--
8	86.67	74.44	--	--

KEY

VAR SET = Variable set

LorT = Longitudinal or transverse action.

C-W-S = Cutting, whittling or scraping.

H-3 = Three-part relative hardness scale for contact materials (see Table 11).

H-5 = Five-part relative hardness scale for contact materials (see Table 11).

Table 18. Dependent variables most highly correlated with the discriminant functions.

IND VARS	VAR SET	VAR HIGH COR DF 1	VAR HIGH COR DF 2	VAR HIGH COR DF 3	VAR HIGH COR DF 4
LorT	8	VTOTFEAT	--	--	--
	16	DTOTSMAL	--	--	--
	32	VFEATHS	--	--	--
C-W-S	8	VTOTFEAT	DTOTFEAT	--	--
	16	DTOTSMAL	VTOTVLRG	--	--
	32	VFEATHS	DHINGEV	--	--
H-3	16	VTOTLARG	VTOTFEAT	--	--
	32	VSTEPM	DFEATHS	--	--
H-5	32	VSNAPV	DHINGES	VHINGES	VSTEPS

KEY

IND VARS = Independent variables which are discriminated.

VAR SET = Variable set (see Tables 10 and 16).

VAR COR HIGH DF n = Dependent variable with highest correlation with discriminant function 1, 2, 3 or 4. The dependent variables are defined on Table 10.

LorT = Longitudinal or transverse action.

C-W-S = Cutting, whittling or scraping.

H-3 = Three-part relative hardness scale for contact materials (see Table 11).

H-5 = Five-part relative hardness scale for contact materials (see Table 11).

Table 19. The percentage of Blind Test tools corrrectly classified using the 16 and 32 variable sets (see Tables 10 and 16).

VAR SET	LvsT	% CORRECTLY CLASSIFIED C-W-S	H-3	H-5
32	100.00	87.50	25.00	12.50
16	100.00	87.50	37.50	0.0

KEY

VAR SET = Variable set

LorT = Longitudinal or transverse action.

C-W-S = Cutting, whittling or scraping.

H-3 = Three-part relative hardness scale for contact materials (see Table 11).

H-5 = Five-part relative hardness scale for contact materials (see Table 11).

The results were respectable for discriminating tools on the basis of the 5-part contact material scales using 32 dependent variables (Table 17). Snap-, hinge- and feather-termination, small, large and very large size, and ventral location are the dependent variables most highly correlated with the DFs (Table 18).

To summarize, the frequency of feather-terminated scars is obviously an important reflection of contact material hardness, as is the frequency of small scars. The location of scars does not appear to contribute significantly to discrimination between contact material classes. The accuracy of discrimination on the basis of contact material was generally lower than discriminating tool actions, but still of an acceptably high level (Table 17). Unfortunately, these promising results are not supported by the blind tests (Table 19). It is apparent that the dependent variables employed are not suitable for discriminating between contact materials using DFA.

8.4.4 *Summary and Discussion of Discriminant Function Analysis Results*

This study has demonstrated that DFA can be employed to accurately classify tools by use-action on the basis of microflake scar variables which can be readily observed with low magnifications. Unfortunately, the same cannot be said of discrimination on the basis of contact materials with the microflake scar variables employed. Stafford (1977:242-244) obtained similarly dissapointing results when attempting to discriminate between contact materials on experimental burins using microflake scar attributes as dependent variables. It is suggested that additional microflake scar variables, such as spacing and configuration (see Chapter 6.5.2) might increase the discriminating power somewhat. A much larger experimental sample, with perhaps 50 tools in each contact material category, would provide a much clearer picture of the suitability of microflake scar variables for reflecting contact material.

Lastly, it should be remembered that the generally promising results of the present study were obtained using combinations of only three microflake scar characteristics on a limited sample of experimental tools. The addition of variables derived from other classes of microwear, such as degree and orientation of edge-rounding, striation orientation and location, and polish characteristics will probably provide very precise discrimination for both tool action and, most importantly, contact material. It is left to the future for such studies to be made as they are beyond the scope of the present exploratory research into multivariate functional analysis.

CHAPTER 9

BLIND TESTS

9.1 Introduction

Microwear research is unique in archaeology in that the accuracy and effectiveness of its methods have been objectively evaluated through a series of "blind tests" (Keeley and Newcomer 1977; Newcomer and Keeley 1979; Odell and Odell-Vereecken 1980; Gendel and Pirnay 1982; Newcomer, Grace and Unger-Hamilton 1986). Although the details differ somewhat from test to test, they all involve one participant studying tools used by someone else and the analyst subjecting his or her functional interpretations to scrutiny and evaluation. In all cases the microwear analyst had no prior knowledge of how or on what the tools were used, or even which part of a tool was used. These tests were primarily undertaken to determine the accuracy of a particular method of microwear analysis, and to evaluate the usefulness of microwear analysis as a research method.

Previous blind test studies were undertaken to evaluate developed methods, and this is where the present series of tests differ. The blind tests discussed in this chapter were undertaken after the microwear traces on the experimental basalt tools had been observed, and a preliminary interpretive framework had been devised for the interpretation of wear traces. The tests were undertaken to point out the strengths and weaknesses in the preliminary conception of the relationship between tool function and microwear patterns, rather than to assess the accuracy of a developed method.

9.2 *Background*

One set of ten basalt tools (Table 20) was used in all four of the blind tests discussed below. These were made and used by Mr. G. Howe, with basalt he collected in 1984 from the same source area as the basalt used in the experiments (see Chapter 3.1, Figures 1 and 2). Howe, currently a graduate student in archaeology at Simon Fraser University, was well qualified for undertaking the tool use part of the blind tests by virtue of his having participated in a previous blind test in a similar capacity (Odell and Odell-Vereecken 1980). In addition, he is a competent flint-knapper and has extensive experience in Canadian Plateau archaeology.

There were few restrictions placed on Howe's use of the blind test tools. Essentially, he was required to stay within the systemic framework discussed in Chapter 4. Howe was thus free to use retouched or unretouched, hafted or unhafted tools made of vitreous or non-vitreous basalt and to use them with any actions or on any contact materials that are historically documented or prehistorically inferred to have been used on the Canadian Plateau within the past 4000 years. This left Howe an almost unlimited amount of latitude for selecting a combination of contact materials and use-actions for each tool. In addition, there were no restrictions on the number of actions each tool could be used with or the number of contact materials each could be used on.

All ten tools were percussion flakes derived from one vitreous basalt cobble. One working edge was retouched prior to use, and several unused areas of another tool were blunted by retouch to facilitate prehension. None of the tools were hafted, although two were held in a leather pad when used.

Howe used the tools in North Vancouver between September 1985 and January 1986 and sent them to me in May 1986 without any clues as to their function. Mr. M. Rousseau, also a graduate student at Simon Fraser University, acted as a middle-man, forwarding Howe's notes on the tools to me after receiving my written interpretations in July 1986.

9.3 *Discussion of Blind Test Results*

9.3.1 *Introduction*

Four blind tests were undertaken using the same set of ten tools: *Blind Test (BT) 1* involved functional interpretations made with only the aid of a hand lens (5 x); *BT 2* employed low magnifications of 40 to 100 x and relied primarily on microflake scar attributes, as well as edge-rounding and polish; *BT 3* employed high magnifications of 200 to 600 x and interpretations were made on the basis of polishes, trend lines, striations and edge rounding; *BT 4* used microflake scar attributes to evaluate the accuracy of a discriminant function analysis program for determining tool function and is briefly discussed in Chapter 8.4. *BT 1* to *BT*

Table 20. Summary of Blind Test tool use.

TOOL #	DESCRIPTION OF USE-EPISODE
1	This tool was deliberately retouched prior to use and the retouched edge (area 1) was used transversely on old, dry mule deer antler (at least 10 years old) (5 min., 200 strokes). The tool was unhafted and held with a bare hand, resulting in small flake removals in area 2. The tool was effective for smoothing the surface of the antler, but was unsuitable for substantially thinning it. The tool was moderately dulled, but still useful.
2	The tool was not deliberately retouched prior to use. Area 1 was used longitudinally on a fresh, moist, red cedar branch contacting bark and wood (15 min., 2000 strokes). The tool was unhafted and held with a bare hand, which resulted in smoothing and polish on the dorsal ridge at area 2. The tool was an effective wood saw and was hardly dulled.
3	The tool was not deliberately retouched prior to use. Area 1 was used transversely to scrape bark and wood from a fresh, moist, red cedar branch (5 min., 500 strokes). The tool was unhafted and held with a bare hand which may have resulted in the microflake removals in area 2. It was effective at scraping bark and wood and became moderately dull, but was still usable.
4	The tool was not deliberately retouched prior to use. Areas 1-4 (the tool tip) were employed with a boring (pushing-twisting) action to perforate a piece of commercially tanned leather which was at least 10 years old (5 mins., 50 strokes). The tip of the tool broke off after a few strokes, but then stabilized. The tool was used unhafted, but the distal end was wrapped with leather to protect the hand. Microflake removals in area 5 are probably the result of prehension. Tool 4 was not particularly useful for perforating the leather, and this became especially difficult after the tip snapped off and the end stabilized.
5	The tool was not deliberately retouched prior to use. Area 1 was used longitudinally on an old, dry mule deer antler beam (at least 10 years old) (5 min., 400 strokes). The tool was held with a bare hand during use. It was useful for producing a shallow groove in the antler beam, but the edge rapidly broke down during use and the tool was ready to be sharpened or discarded after 400 strokes.

continued

Table 20 (continued)

TOOL #	DESCRIPTION OF USE-EPISODE
6	Areas 1 and 6 were deliberately blunted with an antler billet prior to use to remove sharp projections which may have injured the hand. The used areas 2-4 were not retouched prior to use. The tool was used to chop a fresh, moist red cedar branch, contacting both bark and wood (10 min., 500 strokes). Many flakes were removed by the impact during use. It was not really suitable for chopping due to its small size which resulted in substantial hand shock and difficulty in prehension, despite being wrapped in leather.
7	The tool was not retouched prior to use. Area 3 was used longitudinally to slice commercially tanned leather (at least 10 years old) into strips (10 min., 1000 strokes). The tool was held with a bare hand and microflake removals at areas 1 and 2 are probably a resulted of prehension damage. The tool was very effective and barely dulled.
8	The tool was not retouched prior to use. Area 1 was used longitudinally to cut hot roast beef (10 min., 500 strokes). The tool was held with a bare hand. Its small size rendered prehension difficult, but the edge was effective at cutting meat. It was only slightly dulled with use.
9	The tool was not retouched prior to use. Area 1 was used transversely on old, dry mule deer antler (at least 10 years old) (5 min.). The tool was held with a bare hand. As with tool 1, this tool was only capable of smoothing the surface of the antler. It was quite dull, but still usable.
10	The tool was not retouched prior to use. Area 1 was used transversely on a piece of commercially tanned leather (at least 10 years old) (15 min., 900 strokes). The leather rested on a wooden cutting board which was contacted by the tool a few times. The tool was held with a bare hand. The tool was ineffective at scraping leather and accomplished very little. It was moderately dulled but still useful.

3 follow a logical progression from essentially non-microscopic examination of a tool, to low magnification and high magnification study.

The blind tests under discussion consist of four tests undertaken with the same tools and studied by one individual, and are therefore not independent of each other. It is obvious from the way the tests were undertaken that observations made during earlier tests could influence interpretations made in later ones. However, this effect was minimized because the actual functions of the tools were not known until after the completion of all of the tests. In addition, the tests were undertaken at least one week apart and the interpretations made for each test were not consulted during other tests.

It was decided before the tests not to attempt to distinguish between bone and antler as contact materials because of the nearly identical wear patterns produced by working these materials (see Chapter 7.4-7.5). As indicated in Chapter 4, the two materials were used for almost identical technological purposes by the Interior Salish and their ancestors, so very little information would be lost by not distinguishing the materials on prehistoric tools.

Previous blind tests (Odell and Odell-Vereecken 1980) have demonstrated that low magnification methods are not accurate for interpreting specific contact materials, but are quite useful for discerning the relative hardness or resistance of a contact material. Therefore, relative contact material hardness scales were also used in the present series of tests (Tables 21,22). The scales used in the present study are similar to those used previously (see Tringham et al 1974; Odell and Odell-Vereecken 1980).

The manner of use of the ten test tools is described on Table 20, and the interpretations made during the three blind tests are provided on Tables 23-25. Detailed interpretations of the tools are provided in Appendix E. Each blind test is summarized below and the results are discussed, with the purpose of illuminating where the preliminary interpretive framework is useful and where it needs improvement.

9.3.2 *Blind Test 1*

Since this blind test is essentially non-microscopic in nature and is provided as a baseline against which to compare the other tests, it is not necessary to discuss the interpretations (Appendix E) or the results (Table 23) here. The results are discussed in comparison with *BT 2* and *BT 3* in section 9.4. By recording my interpretations of tools made largely on the basis of morphology and flake scar location, I am explicitly admitting my intuitive functional interpretations at the outset. These initial interpretations were treated as functional hypotheses against which the microwear observations in later tests were evaluated. I consider it advantageous to have a "holistic" approach to functional analysis, where non-microscopic tool attributes are considered in concert with microwear traces observed at any magnification (see also Newcomer, Grace and Unger-Hamilton 1986:216).

Table 21. Three-part relative hardness scale for contact materials.

HARDNESS CATEGORY	ACTUAL CONTACT MATERIALS
1. Soft	flesh and fat plants
2. Medium	hide hard and soft woods
3. Hard	bone antler

Table 22. Four-part relative hardness scale for contact materials.

HARDNESS CATEGORY	ACTUAL CONTACT MATERIALS
1. Soft	flesh and fat plants
2. Medium-Soft	hide soft woods
3. Medium-Hard	hard woods
4. Hard	bone antler

Table 23. Blind Test 1 results.
Correct answers are provided in brackets below incorrect estimates.

TOOL #	USED AREA	ACT-ION	CM 3	CM 4	SPECIFIC CM
1	1	T	2 (3)	3 (4)	hard wood (antler)
2	1	L	2	2	hide (wood)
3	1	T	2	2	hide (wood)
4	1-4	B	2	2	hide
5	1	L	3	4	antler/bone
6	1,2	T (C)	3 (2)	4 (2)	antler/bone (wood)
7	3	L	1 (2)	1 (2)	flesh (leather)
8	1	L	1	1	flesh
9	1	T	3	4	antler/bone
10	1	T	2	2	wood (leather)
total correct	10	9	7	7	4

Key

CM=contact material
T=transverse
L=longitudinal
C=chopping
B=boring

9.3.3 *Blind Test 2*

This test involved interpreting tool function primarily on the basis of microflake size, type, configuration and location as well as edge rounding and polish visible at 40 x (and up to 100 x) with a stereomicroscope. The functional assessments made for the test are presented in Appendix E and summarized on Table 24. The results of the test are discussed below.

The used edge or part of the used edge was correctly identified on all tools, and tool action was correctly assessed on 9 out of 10 tools. Tool 6 was incorrectly assessed as having been used with a longitudinal action while it was actually used with a chopping action. Tool 10 was correctly identified as having been used transversely, although it was specifically used with a scraping action rather than whittling, as estimated. There is little reason to distinguish between scraping, shaving, whittling or planing actions as they are all used to thin contact materials. It is apparent that used edge and tool action can usually be identified correctly on basalt tools with the use of low magnifications (see also Odell and Odell-Vereecken 1980).

Accurately identifying specific contact materials solely on the basis of wear patterns visible at low magnifications is not possible (Odell and Odell-Vereecken 1980) and was not a goal of this blind test. On a 3-part relative hardness scale (Table 21), contact material was correctly identified for 7 tools and partially correctly identified for 2 others (Table 24). On a 4-part relative hardness scale (Table 22) contact material was correctly estimated for 5 tools and partially correctly estimated for 4 others (Table 24). Tool 1 was deliberately retouched prior to use and the contact material was estimated as having been medium hard to hard when it actually was hard. The main reason for this error is due to mistakenly identifying some of the microflakes resulting from wear as resulting from retouch. Odell and Odell-Vereecken (1980) caution against over-estimating the hardness of a contact material on retouched tools and the error here is largely a result of over-compensating for this tendency. Tool 2 was used on fresh, damp cedar bark and wood, a material which is classed as medium on the 3-part scale and medium-soft on the 4-part scale (Tables 21, 22). The contact material was estimated as soft to medium on the 3-part scale and soft to medium-soft on the 4-part scale, estimations which are only slightly off. Tools 3 and 10 were estimated to have been used on medium-soft to medium-hard materials on the 4-part scale, while they were both actually used on medium-soft materials. Lastly, the contact material of Tool 9 was estimated as medium and medium-hard on the 3 and 4-part scales, but was actually hard.

In summary, the following "lessons" have been learned from *BT 2*:

1. The used area(s) of a tool can be very accurately identified using low magnifications, even on retouched edges.

Table 24. Blind Test 2 results. Correct answers are provided in brackets below incorrect estimates.

TOOL #	USED AREA	ACTION	CM HARDNESS 3-pt.	4-pt.
1	1	T	2/3 (3)	3/4 (4)
2	1	L	1/2 (2)	1/2 (2)
3	1	T	2	2/3 (2)
4	1-4	B	2	2
5	1	L	3	4
6	1 (1,2)	T (C)	2	2
7	3	L	2	2
8	1	L	1	1
9	1	T	2 (3)	3 (4)
10	1	T	2	2/3 (2)
total correct	9.5	9	8	7

Key

CM=contact material
T=transverse
L=longitudinal
C=chopping
B=boring

2. Tool action can be very accurately assessed using low magnifications.

3. The relative hardness of a contact material can be identified with an acceptable degree of accuracy on both 3- and 4-part hardness scales.

4. Contact material hardness is difficult to interpret on retouched tools.

5. There is a general tendency to slightly underestimate contact material hardness.

9.3.4 *Blind Test 3*

This test involved high magnification (200-600 x) examination of the tools and interpretations based on the wear traces observed (polish, striations, edge rounding) (Appendix E; Table 25). Most observations were made at 300 x using a Unitron Series N metallographic microscope equipped with incident lighting. Tools 4,6,7, and 8 had no visible polish or very poorly developed polish and all lacked striations. If one were interpreting prehistoric tools with similar inconclusive wear patterns, neither tool action or contact material would be inferred. The necessity to come up with interpretations for the test is the only reason they were offered.

Used area(s) and action were correctly identified on all but Tool 6. The most unexpected result of *BT 3*, which consequently allowed the greatest opportunity for refining the methods under development, was the incorrect interpretations of contact material made on the basis of micropolishes. Specific contact material was only correctly identified on four tools. Prior to this test, my preliminary assessment of the experimental data indicated that distinct micropolishes developed from specific contact materials and this belief formed the basis of interpretations of contact materials during the test. Particularly surprising were the incorrect interpretations of Tools 1,9 and 10. The polish on these tools was what was then thought of as "typical" wood polish, although Tools 1 and 9 were actually used on antler and Tool 10 was used on commercially tanned leather. Similar polish was probably not present on experimental antler scraping tools probably because they were not used long enough for such polish to develop. This was no simple oversight; scraping experiments were deliberately cut short due to the extensive damage and dulling of the tool edges and lack of effect on the antler. Tool 9 should have been recognized as having been used on antler due to the extensive edge damage and isolated patches of polish. However, the polish patches were much more extensive than any present on experimental antler scraping tools and the contradictions were rationalized as the tool having been used on very hard, dry wood, even though scraping hard, dry wood in the experiments did not result in exactly the same wear pattern. Scraping commercially tanned leather was not an activity undertaken in the experiments, nor is it one which would have taken place under aboriginal conditions, and it

Table 25. Blind Test 3 results.
Correct answers are provided in brackets below incorrect estimates.

TOOL #	USED AREA	ACTION	SPECIFIC CM
1	1	T	wood (antler)
2	1	L	plant stems (wood)
3	1	T	wood
4	1-4	B	hide
5	1	L	antler/bone
6	4,5 (1,2)	T,G (C)	antler/bone (wood)
7	3	L	flesh (leather)
8	1	L	flesh
9	1	T	wood (antler)
10	1	T	wood (leather)
total correct	9	9	4

Key

CM=contact material
T=transverse
L=longitudinal
C=chopping
B=boring
G=graving

thus violates the conditions of the blind test (see Section 9.2). The bright, smooth, pitted polish on Tool 10 which resulted from scraping the tanned hide was not present on untanned deer hide worked in the experiments, even when dry and hard. It is unlikely that such polish would result from scraping hide under aboriginal conditions.

Incorrect interpretations of contact materials on Tools 6 and 7 were not surprising due to a lack of definite polish on either tool. Contact material simply should not have been inferred from such ambiguous wear traces. The extent of bright, smooth, pitted polish on Tool 2 was very much like distributions of polish found on plant cutting tools, and much more extensive than that found on wood cutting tools used in the experiments. The distribution of polish far from the edge, in particular, was unlike wood working tools, and was the result of deep penetration of the tool into the soft, damp cedar bark.

The following "lessons" were learned from *BT 3*:

1. In cases where only ambiguous traces of polish are present, interpretations of contact material should not be attempted.

2. Micropolishes, trend lines and striations visible at high magnifications are excellent indicators of tool action when present. Unfortunately, these are usually not present on lightly used tools.

3. Longitudinal actions on flesh or hide (leather) result in a very slow development of use-polish.

4. Each contact material does not have a distinctive polish unique to working that substance. The significance of polishes as indicators of contact material must be completely reassessed (see Chapter 10.2.3).

9.4 *Summary and Discussion*

Accurate identification of the used part of a tool and tool action was achieved in all three of the blind tests (Tables 23-25). This result is surprising in that the non-microscopic analysis of *BT 1* resulted in interpretations as accurate, or slightly more accurate, than in the low (*BT 2*) or high magnification (*BT 3*) tests. However, the relative hardness of contact materials was identified slightly more accurately in *BT 2* than *BT 1*, and it must be stressed that the estimates of contact material in *BT 1* were little more than guesses. Specific contact material was identified with equal inaccuracy in *BT 1* and *BT 3*. As already mentioned, contact material was only guessed at in *BT 1*, and a combination of problems, all of which can be overcome, led to the disappointing results in *BT 3*.

CHAPTER 10

RELATIONSHIPS BETWEEN TOOL-USE BEHAVIOUR AND MICROWEAR ON EXPERIMENTAL BASALT TOOLS

10.1 *Introduction*

All of the data pertinent to understanding the relationships between tool use and microwear, including the experimental microwear observations (Chapter 7), the statistical analyses (Chapter 8), and the blind test results (Chapter 9), are synthesized in this chapter. Emphasis is placed on microwear traces visible on vitreous basalt tools with light microscopes, as this part of the research is most likely to be of practical use to others. The non-vitreous basalt and SEM studies are also discussed.

10.2 *Vitreous Basalt Microwear Summary*

In Sections 10.2.1-10.2.3, four types of microwear are discussed as to their usefulness for providing identifications along three principal functional dimensions: (1) the used part of a stone tool, (2) tool action, and (3) contact material. Section 10.2.4 is a discussion of microwear patterns resulting from specific tool-use activities.

10.2.1 *Locating Used Areas*

Before studying microwear traces on a prehistorically used tool, the used part(s) (i.e., edge, point, spur) must be located, since such knowledge, when considered in conjunction with the technological and morphological attributes of a tool can often provide insights into its function. The usefulness of each of the major types of microwear in identifying used edges is discussed below.

Rounding: The presence of substantial edge rounding visible under low magnifications usually indicates a used edge. However, one should be careful not to confuse use-rounding with deliberate grinding of an edge for prehension purposes. Usually, though, edge grinding results in a markedly facetted edge, rather than a rounded one.

Polish: The presence of polish with the characteristics of use-polishes described in Section 10.2.3 is an excellent indicator of a used area. Ambiguous traces of polish should be viewed with caution: these could be the result of platform grinding, deliberate edge grinding, percussor contact, or prehension. Smoothing present on some used edge-aspects lacking polish may also help to identify a used edge.

Striations: When present, striations are almost always found on use-polish, but are often absent on even heavily used tools or visible only at very high magnifications.

Microflake Scars: Almost any type of use will result in microflake removals, but the mistake of confusing use-microflake scars with those resulting from deliberate retouch or incidental edge-damage must be avoided. Deliberate retouch results in contiguous and overlapping scar configurations (see Chapter 6.6.2), while use-retouch scars are sometimes scattered. Unfortunately, incidental damage scars can have either of these distributional characteristics. Regularity of scar size is also unlikely to allow discrimination between deliberately retouched and use-retouched edges, although incidentally damaged edges may have more size variability than either of these two.

Absolute scar size is most likely to be useful for identifying used edges. Most microflake scars resulting from use are smaller than most deliberate retouch scars--scars under .50 mm in length are almost always a result of use, and use scars rarely exceed 2 mm in length. Most retouch scars, on the other hand, are larger than about 3 mm, although smaller ones do occur. Retouch resulting from accidental damage often only occurs on a small portion of a usable edge, unlike most use-retouch, and the scars are usually larger as well.

To summarize, the used part(s) of a tool are readily identifiable by the presence of several types of microwear, and the more types of wear which are present, the more likely one is to correctly identify an area as used. To distinguish "hafting wear", which is actually a form of wear resulting from use, from wear on the edge or projection contacting an objective material, one must consider all of the microwear traces present, in conjunction with the technological and morphological characteristics of the tool, and its hypothesized function.

10.2.2 *Determining Tool Action*

Once the used area(s) of a tool have been isolated, the next step is to determine *how* the tool was used. Tool action should be considered before contact material, because tool action limits the potential contact materials on which the tool may have been used. For example a tool used with a scraping action was unlikely to have been used to work green plants. Each type of microwear is considered below regarding its usefulness in distinguishing between tool actions.

Rounding: Use with a longitudinal action only occasionally produces significant edge rounding. Use on harder materials such as bone and antler sometimes results in localized rounding on blunted parts of the edge, but by the time this rounding occurs the tool is ready to be resharpened or discarded. When present on tools used with longitudinal actions, edge rounding is usually restricted to projecting parts of the edge.

Rounding is common on scraping tools and is sometimes extensive. In addition, the rounding usually extends from the edge onto the dorsal edge-aspect because a scraping action requires most tools (flakes) to be used with the ventral edge leading and the tool tilted so the dorsal aspect contacts the material more. Tools used for whittling have much less rounding. Boring and graving tools have rounded tips and boring tools have additional rounding on edges and ridges back from the tip.

Striations: When present, striations are excellent indicators of tool action--the problem is that they are only visible on about 1/3 of the experimental tools. However, they may be more common on prehistoric tools used under grittier conditions. Longitudinal use results in striations oriented more or less parallel to the edge as well as some at oblique angles to the edge (up to about 50 degrees). Transverse use results in striations roughly at right angles to the edge, with a few at oblique angles (not less than 45 degrees). Striations on the experimental tools are almost always located on polish and one can expect the same with prehistoric tools.

Microflake Scars: The distribution of microflake scars is an excellent indicator of how a tool was used. Scraping actions usually require a tool to be used with the ventral edge-aspect leading, so that most microflake removals are on the dorsal (trailing) aspect. However, hard contact materials such as bone or antler often result in substantial removals from the ventral aspect as well. In general, scars on the ventral aspect are smaller than on the dorsal, scattered rather than clustered and fewer in numbers. In almost every case, the ratio of ventral scars to dorsal scars on scraping tools is less than .5:1 and usually less than .2:1. On wood whittling tools the mean v:d ratio is .5:1, although it is .75:1 for dry wood whittling.

Longitudinally used tools usually exhibit a more even distribution of scars, but in many cases there is a preference for scar removals from the ventral aspect rather than the dorsal. This is probably due to the curvature of most flakes which directs force from the dorsal to ventral surfaces when used longitudinally on a contact material--the more marked the curvature, the more pronounced the preference for microflake scars on the ventral aspect. Scars on tools used with longitudinal actions are usually the same size on both aspects.

Scars often exhibit a "directionality" in the orientation of their long axes. Scars on transversely used tools are mainly oriented at or near right angles to the edge, while those on longitudinally used tools are often oriented at oblique angles to the edge.

The correlations presented in Chapter 8.2 indicate an inverse relationship between snap scars and high spine-plane angled edges. Most of these tools were used for scraping actions, so it is unclear whether it is the action, the spine-plane angle, a combination of both, or some other factors, which are responsible for the paucity of snap scars on scraping tools. Regardless of whether snap scars are related to tool action or spine-plane angle, they are not common on tools used with scraping actions.

Polish: The distribution of polish is useful for determining action. Polish is present on the edge and in roughly equal proportions on both edge-aspects of tools used with longitudinal actions. On scraping tools, most polish is present on the edge and dorsal aspect, with little on the ventral aspect since the edge and dorsal aspects have much more contact with the worked material. Trend lines in polish are sometimes present and indicate the direction of use.

10.2.3 *Identifying Contact Material*

Determination of the material on which a prehistoric tool was used is the most difficult task facing microwear researchers, but contact material identification is very important for reconstructing past human behaviour. It may well be that other methods of contact material identification, such as analysis of residues (e.g., Loy 1983; Anderson 1980; Briuer 1976; Broderick 1982) hold more promise. It is most likely that a combination of chemical and microwear methods will prove to be most effective.

Polish is widely regarded as the most accurate and specific microwear indicator of contact materials. However, other types of wear are also indicative of contact material to a lesser degree of accuracy and specificity and can be used to supplement or modify interpretations made on the basis of polish characteristics.

Rounding: Pronounced rounding extending from the edge onto the dorsal edge-aspect is indicative of hide scraping tools. Wood or bone/antler scraping tools and longitudinally used tools never develop a similar degree of rounding.

Extensive rounding was otherwise only elsewhere observed on tools used for graving or boring bone and antler, instances where limited parts of a tool contact hard materials for extended intervals. In the case of boring, rounding is present on the tip of the tool and additionally on the edges and ridges near the tip. One may expect a similar distribution of rounding on tools used to bore wood and possibly on heavily used hide perforator/borers. Rounding similar to that observed on the bone and antler engraving tools may also be found on tools used to engrave wood.

Striations: Striation frequency probably reflects the amount of grit in the work environment and the ability of the contact material to hold the grit or small fragments of the tool (microflakes) while being worked. Very few striations develop on tools used to work bone and antler which may be due to the inability of these materials to catch and hold the grit or tool fragments which are responsible for striating the tool during use.

Microflake Scars: Microflake scar characteristics and distributions cannot be used to determine specific contact materials, but they are useful for identifying the relative hardness of a contact material (see Chapters 8,9; also Tringham et al. 1974; Odell and Odell-Vereecken 1980; Odell 1981). Feather-terminated scars are believed to be predominant on tools used on soft contact materials and step scars on those used on harder contact materials. Hinge scars are perhaps better indicators of hard materials than step scars, as indicated in Chapter 8.2. If tools used longitudinally are ranked by their percentage of step scars, hide working tools have the most, followed by tools used to work bone/antler, wood, butchering and plants (Table 27). When ranked by hinge scars, bone/antler working tools used longitudinally have the most, followed by tools used on hide, butchering, plants and wood (Table 27). For longitudinally used tools, feather scars predominate on those used to work plants, followed by wood, butchering, hide and bone/antler (Table 27).

Harder contact materials result in larger microflake scars and softer contact materials in smaller scars (see Chapter 8.2). This pattern is reflected by the percentage of small scars of all types observed, with butchering tools having the highest, followed by tools used on hide, plants, wood and bone/antler (Table 27).

On tools used with transverse actions on bone/antler and dry wood there is a higher proportion of ventral (leading) edge-aspect microflake scars than tools used transversely on softer materials.

Large snap scars, which lend a markedly serrated appearance to the edge of a tool have only been observed on bone/antler sawing tools.

Polish: Use-polishes on flints have been touted as the best microwear indicator of contact materials (Keeley 1980; Keeley and Newcomer 1978; Richards 1984). Recently, however, doubts have arisen as to the validity of the simple "contact material = specific polish" model proposed by Keeley (see Newcomer, Grace and Unger-Hamilton 1986). One of the goals of the present research is to determine if polishes develop on Cache Creek basalt tools as a result of use, and

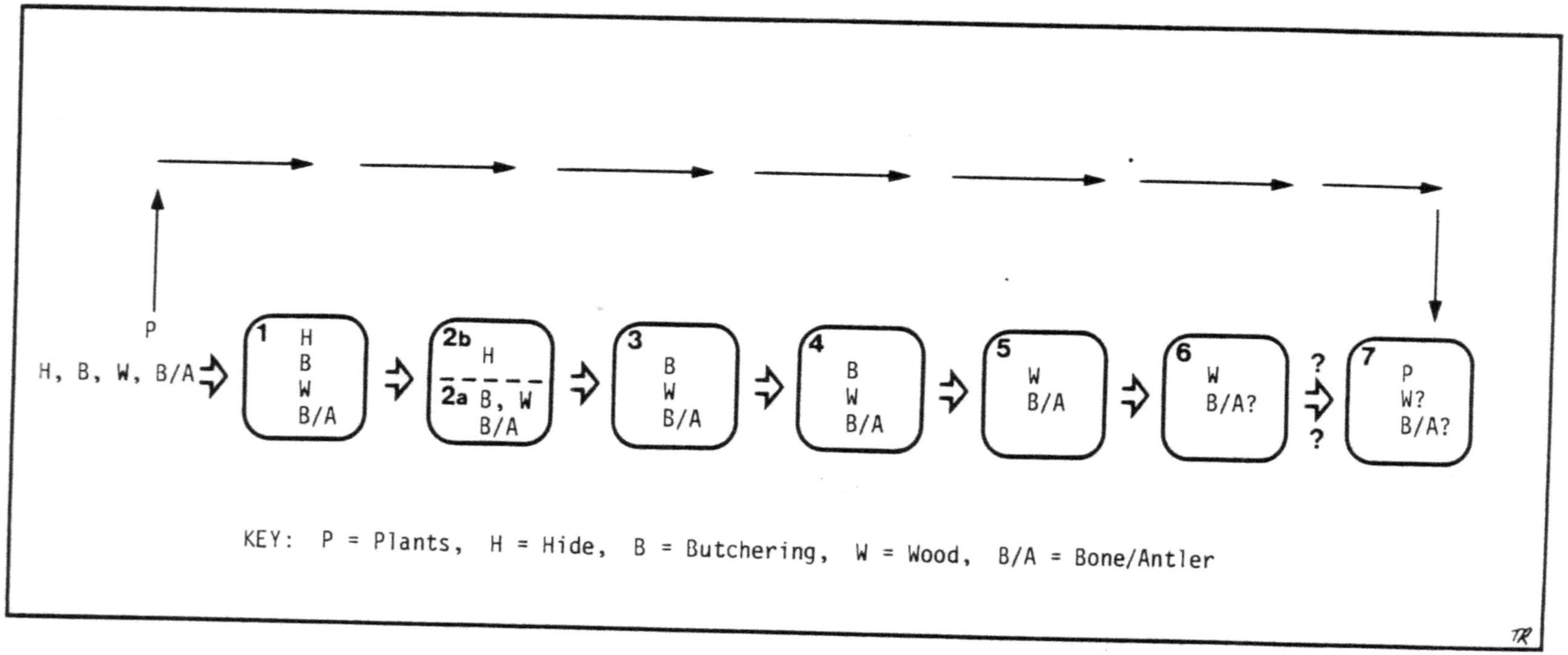

Figure 30. Model of polish stage development on experimental vitreous basalt tools. Polish stages indicated by numbers in the boxes are described on Table 26. Contact materials which result in the polishes are listed in the boxes. The possibility that polish on plant cutting tools does not go through the same stages of development as polish on tools used on other materials is indicated by the line of arrows above the boxes.

Table 26. Descriptions of polish developmental stages on vitreous Cache Creek Basalt tools.

---

1. Initial Undifferentiated Polish Stage: This polish probably results from abrasive "roughening" of the tool surface during early contacts with the worked materials. It is present as scattered traces of dull, rough-textured (grainy-textured) patches on otherwise more reflective surfaces.

2a. Dull to Bright, Grainy Polish Stage: This polish is "grainy" in texture varies in reflectivity from dull to bright, with tiny bright spots or patches surrounded by larger, less reflective areas. It is the second stage of development on tools used for butchering and on bone, antler, wood and possibly plants.

2b. Dull, Grainy Polish Stage: This differs from Initial Undifferentiated Polish primarily in the extent of the polished surface, which can be very large and the "grainyness" of the texture, which is more pronounced. It is generally dull, however, the occasional brighter spot is sometimes present. This polish is found only on hide working tools and is probably an arrested state of development due to continued abrasive surface attrition.

3. Bright to Dull, Grainy Polish Stage: This is much like Stage 2a, but the bright patches are frequently almost as common as dull areas. It is the third stage of development on tools used for butchering and to work bone, antler, wood and possibly plants.

4. Bright to Dull, Grainy Polish/Bright, Smooth Polish Stage: This polish differs from Stage 3 principally by the addition of bright, smooth-textured, unpitted, areas. It is the fourth stage of polish development observed on bone-, antler- and wood-working tools and butchering tools. Polish on plant working tools may also briefly pass through this stage.

5. Bright to Dull, Grainy Polish/Bright, Smooth-Pitted Polish Stage: This polish differs from Stage 4 primarily by the addition of small, shallow pits on the smooth, bright polish. Only polish on tools used to work bone, antler and wood develop to this stage, although polish on plant cutting tools may briefly pass through it.

6. Bright Smooth-Pitted Polish/Bright, Grainy Polish Stage: This stage is much like Stage 5, except grainy-textured polish is rare and bright, smooth-pitted areas predominate. Bright, smooth-pitted polish is generally located on heavily used or projecting parts of the edge and edge-aspects, with bright, grainy polish on the peripheries. Stage 6 polish has been observed on wood-working tools only, although it probably also develops on heavily used bone and antler boring or graving tools. Polish on plant cutting tools may also briefly pass through this stage.

7. Bright, Smooth, Deeply Pitted Stage: Polish is bright to very bright, and smooth textured, but is pockmarked with numerous deep pits. It is very extensive in distribution and has considerable topographic variability. Only polish on plant cutting tools reaches this stage, either passing through the other stages very rapidly or not at all.

---

if so, whether or not distinctive polishes are correlated with specific contact materials. There has been some uncertainty whether polishes useful as functional indicators will develop on relatively granular materials such as basalt (see Odell 1982:25). There is no longer any doubt that polish develops on Cache Creek basalt tools: 85.6% of the vitreous basalt experimental tools had use-polish on them.

A simple model of polish-contact material relationships such as that advocated by Keeley for flint is not appropriate for Cache Creek basalt. Rather, a model is proposed in which general stages of polish development are related to contact materials, tool actions and amount of use. In this model, use-polish on vitreous basalt tools goes through a number of stages of development (Figures 30,31; Table 26) (see Vaughan 1985 for similar observations on Greek flints). However, the rate of development from one stage to another varies depending on the contact material and use-action (Figure 31; Table 27). One result of this fact is that tools become dulled to uselessness when used with some actions on certain contact materials *before the later stages of polish development are reached* (Tables 26,27).

Only one type of polish has been observed on tools used to cut plants and this is bright, smooth-textured and deeply and extensively pitted. It is also much more extensive in area than other smooth polishes observed and has much more topographic variability. The lack of early polish stages on on plant working tools may be due to the possibility that polishes on these tools go through the other stages extremely rapidly.

It is beyond the scope of the present study to investigate the mechanisms of polish formation, however, some speculations are in order. A model developed by Anderson (1980) and modified by Unger-Hamilton (1984) suggests that a very thin layer of amorphous silica may be formed on the surface of flint tools used on plants. The silica is dissolved on the tool surface due to a combination of factors including friction-produced heat, pH levels, and the presence of water and abrasive particles, including plant phytoliths (Anderson 1980:184). Unger-Hamilton (1984:96,97) stresses the importance of chemical reactions *and* abrasion to the process of polish formation. Polish on vitreous basalt tools may develop through similar mechanisms. It is worthwhile noting that tools used to work wood developed polish more similar to plant polish than tools used on any other material, and that silica (phytoliths) is present in both plants and wood (Anderson 1980:187-188). Silica may well be the abrasive agent responsible for the heat build up on the tool surface (especially noted during wood-working experiments in the present study) necessary for dissolution of silica on the tool surface.

Of the other materials contacted in the present study bone and antler working tools are most likely to produce frictional heat during use, although this is probably rapidly dissipated by the constant removal of the surface through microflaking. Unfortunately, when the tool edge stabilizes and heat build-up on the surface becomes possible the tool is often too dull for further use. Polish on bone and antler is especially characterized by highly polished areas being

CONTACT MATERIAL	NUMBER OF USE-STROKES 0	25	50	100	200	300	400	500	750	1000	1500	2000	3000	5000	7000
Hide	-	1	1	2b	2b	2b	2b	2b	2b	2b	2b	2b	2b	2b	?
Butch.	-	1	1	2a	2a	2a	2a	3,4	4	4	4	4	4	?	?
Wood	-	1	2a	2a,3	4	4	4	5	5	5	5	6	6	6	?
Bone/ Antler	-	1	1	2a	3	3	3,4	5	5	5	?	?	?	?	?
Plants	-	1?	2a-6?	7	7	7	7	7	7	7	7	7	?	?	?

Figure 31. Rates of polish development on experimental Cache Creek vitreous basalt tools used on different contact materials. The numbers indicate the stages of polish development described on Table 26. This illustration is somewhat idealized and does not include the variable of tool action which can determine the polish stage reached.

restricted to projecting points of the edge which stabilized earlier than surrounding areas and which were thus subjected to longer use.

Hide was an abrasive, but relatively soft material. Hide working did not cause an excessive build-up of heat on the tool surface, probably due in part to the slow rate of work, the extensive area of a tool contacting the hide, and the constant abrasive removal of material from the tool surface. Instability caused by the continual removal of the tool surface prevented polish from developing beyond a rudimentary stage (Stage 2b, Table 26).

10.2.4 *Specific Activity Microwear Patterns*

The following activities are discussed in terms of their resulting distinctive use-wear characteristics or patterns. The wear descriptions are for tools which have been moderately to heavily used and ready for resharpening or discarding, which is the point where the most distinctive wear has developed on them (Table 27).

Plant Cutting: Plants are one of the softest contact materials on which longitudinally used tools were employed. Dry plants are notably harder than green plants, but are still soft. Microflake scars are relatively evenly distributed on dorsal and ventral edge-aspects, with a tendency to be more common on the ventral aspect. Edges are only slightly rounded, even after heavy use. Striations are rare to common, parallel or oblique to the edge, and short. Stage 7 polish (see Section 10.2.3) develops quickly and covers extensive areas of the edge and both aspects after only moderate use.

Butchering: This would have been a very common activity in prehistoric times as 50-65% of the protein consumed in the last 1500 years in the study area was derived from anadromous fish (Chisolm 1986:115-124) and an estimated 20-40% consisted of deer, elk, mountain sheep and small mammals (Richards and Rousseau 1987). Simple sharp flakes are very effective for butchering fish or deer, including the activities of evisceration, skinning, dismembering, filleting and defleshing. Tools can be used for extended periods without dulling, and this slow rate of dulling coincides with a slow rate of wear development, often resulting in wear traces being difficult to detect, even on heavily used tools. Microflake scars are fairly evenly distributed, but tend to be somewhat more common on ventral aspects. Edges are unrounded to slightly rounded. Stage 4 polish is present, but never extensive, in distribution. Striations are only present on a few tools and are rare on these; they are parallel to the edges and variable in length.

Hide Scraping: Hafted or unhafted basalt tools are effective for defleshing or dehairing deer hide. The fresh, flesh side of a hide is the softest contact material scraped, while the hair side is a harder material than wood in several microflake categories believed to reflect the hardness of a contact material. On most tools microflake scars are much more common on the dorsal (trailing) aspect than ventral (leading) aspect. However, on a few tools this proportion is reversed,

Table 27. Summary of microwear patterns on experimental Cache Creek basalt tools.

	CONTACT MATERIAL	POLISH Stage	POLISH Distri-: bution	ROUNDING :Amount	ROUNDING Orien-: tation:	STRIAE: Orien-: tation:	MICROFLAKE SCARS %Feath Scars	MICROFLAKE SCARS %Small Scars	MICROFLAKE SCARS V:D Scar Ratio
LONG ACTION	Plants	7[a]	V-E-D[b]	S[c]	E[d]	P,O[e]	52.9	82.4	1.2
	Butch.	4	V-E-D	S	E	P,O	45.1	96.0	1.6
	Wood	6	V-E-D	S	E	P,O	46.0	80.0	.8
	Hide	2b	V-E-D	S	E	P,O	36.8	83.0	.6
	Bone/ Antler	3/4/5	V-E-D	S	E	-	26.2	72.1	1.6
TRANS ACTION	Wood: S[f]	6	E-D	M-P	E-D	R	41.4	92.6	.2
	W	6	E-D	S-M	E-D	R	43.2	89.3	.5
	Hide	2b	E-D	P	E-D	R	39.8	86.6	.6
	Butch.	2a	E-D	A	-	R	24.3	74.1	.4
	Bone/ Antler	5	E-D	A-S	E-D	R	22.8	72.6	.6

KEY:

a. see Table 26 for polish stage definitions

b. V = Ventral edge aspect
E = Edge
D = Dorsal edge aspect

c. S = Slight rounding
M = Moderate rounding
P = Pronounced rounding
A = Rounding Absent

d. E = Edge
D = Dorsal

e. P = Parallel to edge
O = Oblique to edge
R = Right angle to edge

f. S = Scraping
W = Whittling

although the tools were used in identical fashion. It is believed that massive edge rounding has obliterated most of the scars on the dorsal edge-aspects of these tools. Stage 2b polish appears after very little use and only becomes slightly brighter after extremely prolonged use. Striations are rare and almost always oriented at or near right angles to the edge.

Cutting Hide: Basalt flake tools are effective for cutting deer hide. Fresh hide encountered while skinning a deer is a relatively soft contact material, but dry, stiff hide is much harder. Skinning is discussed under "Butchering"; dry hide cutting is the focus here. Microflake scars tend to be more common on the dorsal edge-aspect than on the ventral. Edges are only slightly rounded and Stage 2b polish is extensive on both aspects. Striations are common, variable in length and are oriented parallel or obliquely to the edge.

Sawing Wood: Using a back and forth sawing action is an effective way to cut through branches, especially if one wants a "clean" break, a desirable goal when making arrow or spear shafts. Once a groove is established, the shaft can be snapped quite cleanly at that point. Wood is clearly a medium contact material, softer than bone/antler, roughly as hard as dry hide, but harder than fresh hide, plants or flesh. Microflake scars are fairly evenly distributed on dorsal and ventral edge-aspects. Stage 6 polish is present on the slightly rounded edges and on both aspects. Striations are very common on one experimental tool and are oriented parallel or oblique to the edge.

Whittling Wood: Whittling is an action used to thin or sharpen wood. In this series of experiments bone or antler has proven impossible to whittle with basalt tools. In all cases, the ventral edge-aspect contacted the wood and on all but one tool there are fewer microflake scars on the ventral than dorsal aspects. Microflake scars are scattered or clustered in small groups on the ventral aspects and are more or less continuous on the dorsal aspects. Edge rounding or blunting is slight to moderate. Stage 6 polish is extensive, but not as extensive as on wood scraping tools with their wider rounded edges. Most polish is present on the edge and dorsal edge-aspect, with variable, but usually lesser, amounts on the ventral aspect. Striations are rare to common and are oriented at or near right angles to the edge. Wear patterns are very similar to those on tools used to scrape wood, however, the spine-plane angles on the scraping and whittling tools are substantially different. A problem in assessing the suitability of certain spine-plane angles for specific actions is that tools with spine-plane angles pre-judged to be appropriate for scraping or whittling were selected for use in the experiments. The spine-plane angles of tools used for whittling have a mean of 29.7 degrees and ranged from 10 to 50 degrees, while scraping tools had a mean angle of 56.3 degrees and ranged from 40 to 70 degrees. It was observed during the experiments that low-angled scraping tools were not as effective or easy to use as high-angled ones and that whittling tools with higher angles were not as effective or easy to use as tools with lower spine-plane angles. One may speculate that prehistoric wood whittling tools will have spine-plane angles of less than 50 degrees and wood scraping tools to have spine-plane angles generally higher than 50 degrees.

Wood Scraping: Basalt tools are effective at removing bark and thinning branches--activities associated with manufacturing implement shafts (e.g. spear, arrow, harpoon shafts, etc). Edge rounding is moderate to pronounced, and oriented to the dorsal (trailing) aspect, but it is never as pronounced as on hide scraping tools. Stage 6 polish is extensive on the rounded edges, with trend lines oriented at right angles to the edge sometimes visible. Striations are rare to common, variable in length, and oriented at or near right angles to the edge. While almost identical to wear on wood whittling tools, the higher spine-plane angle of scraping tools should serve to distinguish between the two actions.

Sawing Bone/Antler: Basalt tools can be used for sawing into bone or antler, but they suffer such extensive damage after little use that they cannot realistically be viewed as a means of cutting through these materials. The tools are useful for cutting shallow grooves into bone or antler, suitable perhaps for decorative purposes, but little else. It would take dozens of tools to cut through an antler beam, a task which would be much more easily accomplished with a cobble chopper (evidence for "chopping" marks have been observed on antler beams found in prehistoric sites). Most shaping of bone and antler tools was probably accomplished with sand- or silt-stone abraders, which are found in late prehistoric sites. It is also worth noting that crypto-crystalline lithic materials, which are harder than basalt, may have been more effective at working bone and antler. Bone and antler were the hardest materials on which tools were used with longitudinal actions, a fact which is clearly reflected in the microflake scar characteristics (Table 33). Edge rounding or blunting is absent to slightly present. Polish is bifacial and varies from traces to scattered small patches, primarily on high points. It usually only develops to Stage 3 because of massive early damage, although occasionally Stages 4 and 5 are reached. Striations are absent.

Scraping Bone/Antler: Basalt tools used to scrape bone or antler accomplish little beyond a slight smoothing of the contact surface while the tools suffer extensive damage and dulling. Again, this is not an effective way to modify bone or antler, and it is suggested that major alterations were made by using choppers, cryptocrystalline lithics and sand- or siltstone abraders. Microflake scars clearly reflect the hardness of the contact materials (Table 27). Microflake scars are scattered to absent and predominantly small in size on the ventral (leading) aspect; on the dorsal (trailing) aspect they are continuously distributed along the used edge and much more variable in size. Edge rounding only occurs on a few isolated projections and is rarely more than slight. Experimental tools used until they were very dull have Stage 3 and Stage 4 polish on them, although one retouched blind test tool had Stage 5 polish on a projecting part of the edge. Polish varies from the slightest traces to extensive patches on the edge and dorsal edge-aspect; it is virtually absent from the ventral edge-aspect. One tool has a single striation--it is short, and oriented at approximate right angles to the edge.

Boring Antler: Spurred basalt tools are effective for boring into antler (and presumably bone), especially soaked antler. The edges and tip are extensively rounded with Stage 6 polish concentrated on the rounded loci. Trend lines are present on polish on the edges and ridges and run at or near right angles to the long axis of the boring tip. Striations are absent.

Graving Bone/Antler: Graving is a useful action to provide blanks for making tools through the "groove and splinter" technique as the experiments indicated. Graving tools are extensively rounded on the tips and nearby edges and ridges. Stage 6 polish is extensive and concentrated on the rounded areas. Present in the polish on the bone graver are numerous comet-shaped pits which parallel the direction of tool motion. Striations are absent.

10.3 *Non-Vitreous Basalt Microwear Summary*

Nineteen non-vitreous basalt tools were used with scraping, whittling, slicing and sawing actions for butchering deer and fish and for working hide, bone, antler and wood. The amount of use ranged from 165 to 1600 strokes (1:09 to 14:26 minutes). The microwear was observed with a low magnification stereomicroscope and high magnification incident light microscope as described in Chapter 6.2 to 6.5. Used edges were not studied with the aid of a SEM. Non-vitreous tools were also not used in the blind tests.

Results of the non-vitreous basalt microwear study are summarized below:

1. Microflake scars are difficult to distinguish on most tools and their characteristics are hard to clearly observe.

2. Microflake scars are less common than on vitreous basalt tools used for the same tasks and for similar numbers of strokes.

3. Polish is commonly present (at least traces of polish on 63.2% of the tools), although it does not seem to vary in brightness or texture according to contact material, tool action or amount of use.

4. Edge rounding is very common, extensive and occurs after relatively little use regardless of contact material or action.

5. Striations were not observed.

The significantly different microwear characteristics of vitreous and non-vitreous basalts should not be surprising, especially when the differences in their composition are considered (compare Figures 5 and 6 with Figures 11 to 14). It is suggested here that the fine-grained, relatively homogenous, vitreous basalt generally acts much like an isotropic substance when subjected to wear-producing forces (see Lawn and Marshall 1979:66). Medium- or coarse-grained basalts with their more heterogenous composition of numerous large phenocrysts are more like anisotropic solids (Lawn and Marshall 1979:66) and therefore tend to wear much differently. The following rather speculative explanations are offered for the peculiarities of the microwear characteristics of non-vitreous basalts:

1. Force which would cause microfractures on vitreous basalt may be dissipated by being deflected around the numerous large phenocrysts (Lawn and Marshall 1979:66), resulting in ambiguously defined fracture scars (microflake scars) and generally fewer scars overall.

2. Abrasive contact may rapidly wear down the edges due to the large phenocrysts being easily broken off or "plucked" out of the matrix (Greiser and Sheets 1979:292). This rapid wearing down of the edge seems to result from contact with all materials worked in the experiments. The surfaces of the rounded edges vary little in appearance regardless of the contact material and this may be a result of the propensity of the rather soft non-vitreous basalt surface to be continually abraded without stabilizing and becoming "polished".

3. The abraded, used surface is simply too coarse and uneven for striations to be visible.

In conclusion, non-vitreous basalt does not share use-wear characteristics with vitreous basalt, and the potential for deriving suitable models for determining prehistoric tool function from the wear patterns observed on these experimental tools is much more limited. Used edges and tool action would be identifiable on prehistoric non-vitreous tools, but the determination of contact material on the basis of microwear traces alone is not possible with the present limited experimental data.

10.4 *Scanning Electron Microscope Study Summary*

Edge segments of 26 vitreous basalt tools were studied with the aid of a SEM and it is worthwhile briefly examining the results, especially in comparison with observations made with light microscopes. The SEM sample consists of edge segments of 23 used tools and edge segments of 3 unused tools prepared and studied using methods described in Chapter 6.2-6.4. The used tools were employed with scraping, whittling, sawing and slicing actions on wood, plants, hide, bone, antler and for butchering deer and fish. They were used for 100 to 5200 strokes (0:55 to 58:04 minutes) and are mostly heavily used tools selected to maximize the possibility of observing wear with the SEM.

Polish was present on fifteen tool segments (65.2% of the used segments examined); eight of these also had smoothing of the edge-aspect surface beyond the polish and two others had smoothing but no polish. In no case was polish observed with the SEM on tools where it was not observed with a light microscope, however, in three cases definite polish was observed with the SEM, while only traces of polish were observed with a light microscope. Another six segments examined with the SEM had no observable polish, yet polish was observed with a light microscope. The latter six tools were used on bone or antler, and generally only had small amounts of grainy polish visible with a light microscope. Small amounts of grainy polish visible on butchering tools under a light microscope were also not observed with the SEM. Instead, smooth unpitted polish was commonly observed with the SEM, this being only sporadically observed with the light microscope.

Micropolish characteristics have some correlations with contact materials, although the important attribute of "brightness" cannot be observed with a SEM. Under light microscopes, polish on hide working tools is grainy-textured and dull in intensity (Figures 37-45). The SEM study clearly demonstrates that this polish is distinctly different from that found on tools used to work other materials. The polish is comprised of small, overlapping "plates", which are themselves smooth-textured (Figures 77-79, 82,83). The surface appears more like one which is abraded rather than polished. On the hide cutting tool examined, which was used on drier, stiffer hide than the scraping tools, the polish has a "platy" texture around the peripheries and a smoother texture near the edge (Figures 84,85). This smooth polish was not observed with the light microscope.

On the one plant cutting tool examined, polish is extremely extensive, smooth textured and deeply pitted (Figures 72-74), much as it appears under high magnifications with a light microscope (Figures 34-36). It also has the appearance of being built up on the tool surface. Polish on tools used to work wood is much less extensive in distribution, smooth-textured, unpitted and has a "smeared on" appearance (Figures 114, 116, 118, 119). This contrasts with observations made with light microscopes where the polish often appears pitted (Figures 66-69), but not nearly to the extent observed on tools used to cut plants. These shallow pits are not visible as such with the SEM.

Polish is sporadically present on butchering tools and is characteristically smooth-textured and only rarely pitted (Figures 98, 99, 101-107). It lacks the smeared on appearance of polish on wood working tools and appears more like an abrasive smoothing. Definite polish was not observed on any of the tools used to work bone or antler, although the heavily used boring and graving tools were not examined.

Microflake scars are present on all used edges and rounding is very common, although it is sometimes only visible at very high magnifications. Eleven segments had definite striations and a further four had possible striations. Often, striations are only observed at extremely high magnifications and in all cases they are located on polish.

In conclusion, the SEM study served to confirm the distinctiveness of polish developed by working fresh hides. This polish becomes less distinct as the hide becomes drier, stiffer and harder, and becomes more like smooth, unpitted polish observed on butchering tools. Polish developed on plant cutting tools and wood working tools appears "built up" on the tool surface, although this is possibly just an effect of the dissolution of the tool surface and formation of a thin layer of amorphous silica (Anderson 1980; Unger-Hamilton 1984).

CHAPTER 11

SUMMARY AND CONCLUSIONS

11.1 *Summary of Research Conclusions*

As stated in Chapter 1, this study was undertaken to provide a framework for the functional interpretation of tools found at late prehistoric sites located in the southwestern interior of British Columbia. Chipped stone tools were manufactured from vitreous and non-vitreous basalts for use in controlled experiments replicating late prehistoric-early historic tool-use behaviour of Interior Salish peoples indigenous to the study area. Over one hundred tools were used on hide, flesh, plants, wood, bone and antler with slicing, sawing, scraping, whittling, boring and graving actions.

The basalt used in this study came from lower Cache Creek, on the periphery of the Arrowstone Hills in the southwestern interior of British Columbia. It was collected in the form of cobbles from secondary glacial and alluvial deposits. Thin section analysis of 17 cobbles and comparison with published petrographic descriptions of basalts in the area allows tentative assignment of the samples to known formations. The vitreous basalt was primarily derived from the *Chilcotin Group* (ca. 2 to 3 and 6 to 10 million years) and the non-vitreous basalt from the *Kamloops Group* (ca. 42 to 52 millon years). It is known from the work of the 19th century geologist-ethnographer George Dawson and others that glassy basalt from the vicintiy of the Arrowstone Hills was the primary source of lithic raw materials used by Native people in this area.

Microflake scar attributes from 90 vitreous basalt tools were subjected to a series of statistical analyses. Spearman Rank Order Correlations between dependent and independent variables indicated several weak relationships. Spine-plane angle is negatively correlated with snap-terminated scars and positively correlated with step-terminated scars. Contact material hardness is negatively correlated with small microflake scars and positively correlated with larger scars. Lastly, contact material hardness is negatively correlated with feather-terminated microflake scars and positively correlated with hinge-terminated scars.

Cluster analysis was used to group the tools using microflake scar attributes as dependent variables. Complete linkage proved to be the best agglomeration method and cosine the best distance measure. The resulting clusters clearly grouped tools on the basis of use-action and only secondarily by contact material. The same attributes were used for discriminant function analysis, with similar results: discrimination of use-action was achieved, acceptably accurate discrimination of contact material was not. The statistical analyses indicate that microflake scar attributes alone do not sufficiently reflect contact materials to allow accurate functional classifications on the basis of contact material to be made. These attributes do, however, strongly reflect the action with which a tool was used.

Blind tests were used to point out strengths and weaknesses in a preliminary, intuitive interpretive framework. Tool actions were accurately interpreted using both low and high magnifications, as was relative contact material hardness on the basis of observations made at low magnifications. However, exact contact material was not accurately interpreted on the basis of polish characteristics observed at high magnifications. An important result of the blind tests was the rejection of a model relating each contact material class to one particular type of polish.

Micropolish on vitreous basalt tools was found to go through a series of developmental stages, each with its own distinctive characteristics. Contact material, duration of use, and use-action are the three variables which determine the polish stage developed on a used tool. When the polish stage is considered along with edge-rounding, microflake scar characteristics and striations, tool action and contact material can be reconstructed accurately. In particular, plant cutting, butchering, hide scraping, cutting hide, sawing, whittling, and scraping wood, and sawing, scraping, boring and graving bone/antler leave distinctive wear patterns on used tools.

Seven specific problems were listed in Chapter 1 and have been dealt with throughout this study. The problems are listed below, along with a summary of how each has been addressed:

1. *Do vitreous and non-vitreous basalts have similar microwear characteristics?*

Non-vitreous basalt differs from vitreous basalt in its use-wear characteristics (see Chapter 10.3). Most notably, microflake scars are less common and less distinctive, polish is present less frequently and it does not seem to vary in brightness or texture, edge-rounding is common and extensive, and striations were not observed.

2. *Which types of microwear are most useful for identifying the used area of a tool, tool action and contact material?*

As discussed in Chapter 10.2, several wear types are useful for identifying the used area of a tool, although edge rounding and polish are most informative. Tool action can be most reliably inferred from striation orientation, however, striations are only found on about 1/3 of the experimental tools. Microflake scar characteristics also accurately reflect tool action, and additionally, they are usually common on a used edge and easy to observe. The relative hardness of a contact material can be identified with an acceptable degree of accuracy using microflake scar characteristics, edge-rounding and polish visible at low magnifications (see Chapter 9.3.3). Polish characteristics visible at high magnifications are more specific indicators of contact material, although the most useful results will be obtained by considering all classes of microwear (see Chapter 10.2.3).

3. *Which magnifications are most appropriate for functional interpretations of basalt tools?*

Low magnifications are suitable for interpreting tool action and relative contact material hardness. More specific interpretations require the additional use of high magnifications, primarily in the 200 x to 300 x range, but up to at least 400 x. Each of the four classes of wear provide the most functional information when viewed at certain optimum magnifications. Low magnifications of around 50 x are suitable for observing edge rounding and microflake scars, while polishes are best viewed at 200 x to 300 x and striations from 200 x to 2000 x.

4. *Do distinctive types of polish develop on basalt tools as a result of use on different contact materials as has been found to be the case with various flints and cherts?*

A simple model relating a specific contact material to a specific polish (see Keeley 1980) is not appropriate for Cache Creek basalt. Instead, a model relating general stages of polish development to contact material, tool action and duration of use is proposed (see Chapter 10.2.3).

5. *Are rarely observed wear types such as polish and striations more commonly observed when high magnifications are employed?*

Use-polish characteristics and striations are only clearly visible at magnifications of 200 x to 300 x using light microscopes, and magnifications of 400 x to 600 x are often required to identify striations. Extremely high magnifications (above 1000 x) attained with the SEM have revealed striations which were not observable with light microscopes (see Chapter 10.4).

6. *Are multivariate statistical techniques appropriate for the functional classification of basalt tools (and stone tools in general), and can statistical methods be used to identify functionally significant wear variables?*

Both of these questions can be answered in the affirmative and have been discussed above.

11.2 *Future Research Directions*

The present study should not be considered a final statement regarding the relationships between tool-use behaviour and microwear on Cache Creek basalt tools. Rather, it is a significant step towards understanding these relationships and allows the recognition of research directions necessary for deeper understanding. Several recommended areas for future research are listed below:

1. A large sample of prehistoric basalt tools from *Plateau Pithouse tradition* sites in the southwestern interior of British Columbia should be examined for microwear traces. It will be especially important to identify any microwear traces not compatible with observations made on the experimental tools. Further experiments will be necessary to identify the origin of any new types of wear. Examination of a limited number of prehistoric tools by the author has not resulted in the discovery of unexpected microwear traces.

2. A set of experiments similar to the present study should be undertaken with retouched tools to determine if the microwear characteristics are similar to the unretouched tools. Examination of the few retouched tools used in the present study suggests that few differences can be expected, however, much can be learned about distinguishing use-related microflake scars from deliberate retouch scars.

3. Tool actions not employed in the present study should be experimented with. These include drilling, chopping and projectile use.

4. More experiments with hafted tools should be undertaken to determine if hafting has an effect on the development of microwear. Examination of the few hafted tools used in the present study suggest no significant differences exist, however, only hide scraping tools were hafted and the effects of hafting on tools

used with other actions and on different contact materials are unknown.

5. The potential of multivariate statistical methods for functional classification has been illustrated, but much more work remains to be done. Discriminant function analysis holds particular promise, especially with dependent variables derived from additional classes of microwear.

6. At present, the enormous potential of the SEM cannot be applied to the analysis of archaeological specimens because of the necessity of sawing off small segments of a tool and coating it with gold. Efforts should be directed towards developing a material for obtaining casts of used tool edges suitable for detailed analysis with a SEM. Some advances in this area have produced casts of used tool edges accurate enough for the study of use-polish only visible at high magnifications with light microscopes (e.g. Plisson 1983; Knuttson and Hope 1984).

7. Although not microwear in the traditional sense, organic residues on used tools should be investigated. Data obtained from residues and microwear could allow very accurate and specific identifications of contact materials to be made.

REFERENCES

Afifi, A.A. and Virginia Clark
1984 *Computer-Aided Multivariate Analysis.* Lifetime Learning Publications, Belmont, California.

Anderson, Patricia C.
1980 A testimony of prehistoric tasks: diagnostic residues on stone tool working edges. *World Archaeology* 12(2):181-194.

Bamforth, Douglas R.
1986 A comment on "Functional variability in an assemblage of endscrapers". *Lithic Technology* 15 (2): 61-64.

Bevier, M.L.
1982 *Geology and Petrogenesis of Mio-Pliocene Chilcotin Group Basalts, British Columbia.* Ph.D. thesis, University of California, Santa Barbara.

1983a Regional stratigraphy and age of Chilcotin Group basalts, south-central British Columbia. *Canadian Journal of Earth Sciences* 20 (4): 515-524.

1983b Implications of chemical and isotopic composition for petrogenesis of Chilcotin Group basalts, British Columbia. *Journal of Petrology,* 24 (2): 207-226.

Binford, Lewis R.
1981 *Bones: Ancient Men and Modern Myths.* Academic Press, New York.

1982 Objectivity--explanation--archaeology--1981. In *Theory and Explanation in Archaeology*, edited by C. Renfrew, M.J. Rowlands and B.A. Segraves, pp. 125-138. Academic Press, New York.

Boas, Franz
1891 The Shuswap. *Report of the British Association for the Advancement of Science* 1886-1889: 632-647.

Bouchard, Randy and Dorothy I.D. Kennedy
1977 *Lillooet Stories.* Sound Heritage 6 (1).

1979 *Shuswap Stories.* Commcept Publishing Ltd., Vancouver.

Brink, John W.
1978a *An Experimental Study of Microwear Formation on Endscrapers.* National Museum of Man, Mercury Series, Archaeological Survey of Canada, Paper No. 83.

1978b The role of abrasives in the formation of lithic use-wear. *Journal of Archaeological Science* 5: 363-371.

Briuer, Frederick L.
1976 New clues to stone tool function: plant and animal residues. *American Antiquity* 41 (4): 478-484.

Broadbent, Noel D. and Kjel Knutsson
1975 An experimental analysis of quartz scrapers. Results and applications. *Fornvannen* 70: 113-128.

Broderick, Michael
1979 Ascending paper chromatographic technique in archaeology. In *Lithic Use-Wear Analysis,* edited by B. Hayden, pp. 375-383. Academic Press, New York.

1982 Residue analysis of artifacts and flakes recovered from the Telep site DhRp 35. Appendix 2 in *The Telep Site: A Late Autumn Fish Camp of the Locarno Beach Culture Type*, by W.R.B. Peacock. Report on file, Heritage Conservation Branch, Victoria, B.C.

Brose, David S.
1975 Functional analysis of stone tools: a cautionary note on the role of animal fats. *American Antiquity* 40 (1): 86-94.

Buller, Hans
1983 Methodological problems in the microwear analysis of tools selected from the Natufian sites of El Wad and Ain Mallaha. In *Traces d'Utilisation sur les Outils Neolithiques du Proche Orient,* edited by M. C. Cauvin, pp. 107-125. Travaux de la Maison de l'Orient No. 5. Maison de l'Orient, Lyon.

Burton, John
1980 Making sense of waste flakes: new methods for investigating the technology and economics behind chipped stone assemblages. *Journal of Archaeological Science* 7: 131-148.

Cahen, D., L.H. Keeley, and F.L. Van Noten
1979 Stone tools, toolkits, and human behaviour in prehistory. *Current Anthropology* 20: 661-683.

Campbell, R.B. and H.W. Tipper
1971 *Geology of Bonaparte Lake Map-Area, British Columbia.* Geological Survey of Canada, Memoir 363.

Carlson, Catherine
1980 Excavations at the Curr site. In *The Archaeology of Kamloops* by R.L. Wilson and C. Carlson, pp. 87-123. Department of Archaeology, Simon Fraser University, Publication 7.

Chandler, Susan M. and John A. Ware
1976 The identification of technological variability through experimental replication and empirical multivariate analysis. *Newsletter of Lithic Technology* 5 (3): 24-26.

Chisholm, Brian S.
1986 *Reconstruction of Prehistoric Diet in British Columbia Using Stable-Carbon Isotopic Analysis.* Ph.D. thesis, Simon Fraser University, Burnaby.

Corruccini, Jo Ann
1985 Moisture and the formation of obsidian striations. *Lithic Technology* 14 (1): 33-35.

Cotterell, Brian, B. Hayden, J. Kamminga, M. Kleindienst, R. Knudson and R. Lawrence
1979 The Ho Ho classification and nomemclature committee report. In *Lithic Use-Wear Analysis*, edited by B. Hayden, pp. 133-135. Academic Press, New York.

Cotterell, Brian and Johan Kamminga
1979 The mechanics of flaking. In *Lithic Use-wear Analysis*, edited by B. Hayden, pp. 97-112. Academic Press, New York.

Dawson, George M.
1891 Notes on the Shuswap people. *Proceedings and Transactions of the Royal Society of Canada* 9: 3-44.

1896 Report on the area of the Kamloops map-sheet. *Geological Survey of Canada, Annual Report for 1894, New Series,* Vol. VII, Pt. B., pp. 1-348.

Del Bene, Terry A.
1979 Once upon a striation: current models of striation and polish formation. In *Lithic Use-Wear Analysis,* edited by B. Hayden, pp. 167-177. Academic Press, New York.

Del Bene, Terry A., and Phillip H. Shelley
1979 Soapstone modification and its effects on lithic implements. In *Lithic Use-Wear Analysis,* edited by B. Hayden, pp. 243-257. Academic Press, New York.

Diamond, George
1979 The nature of so-called polished surfaces on stone artifacts. In *Lithic Use-Wear Analysis,* edited by B. Hayden, pp. 159-166. Academic Press, New York.

Doran, J.E. and F.R. Hodson
1975 *Mathematics and Computers in Archaeology.* Edinburgh University Press.

Drysdale, C.W.
1914 Geology of the Thompson River Valley below Kamloops Lake, B.C. *Geological Survey of Canada, Summary Report* 1912, pp. 115-150.

Duff, Wilson
1964 *The Indian History of British Columbia. Volume I: The Impact of the White Man.* Anthropology in British Columbia, Memoir 5. British Columbia Provincial Museum, Victoria.

Duffell, S. and K.C. McTaggart
1951 *Ashcroft Map-Area, British Columbia.* Geological Survey of Canada, Memoir 262.

Dumont, John V.
1982 The quantification of microwear traces: a new use for interferometry. *World Archaeology* 14(2): 206-217.

Eldridge, Morley and Arnoud H. Stryd
1983 *C.N. Rail Railyard Expansion Project Heritage Mitigation Study Kamloops Junction, B.C.* Report on file, Heritage Conservation Branch, Victoria.

Ewing, Thomas E.
1981a Regional stratigraphy and structural setting the Kamloops Group, south-central British Columbia. *Canadian Journal of Earth Sciences* 18: 1464-1477.

Ewing, Thomas E.
1981b Petrology and geochemistry of the Kamloops Group volcanics, British Columbia. *Canadian Journal of Earth Sciences,* 18: 1478-1491.

Fedje, Daryl
1979 Scanning electron microscopy analysis of use- striae. In *Lithic Use-Wear Analysis,* edited by B. Hayden, pp. 179-187. Academic Press, New York.

Ferrier, W.F.
1896 Petrographical characters of some rocks from the area of the Kamloops map-sheet, British Columbia. *Geological Survey of Canada, Annual Report for 1894, New Series,* Vol. VII, Part B, pp. 349-427.

Flenniken, J. Jeffrey, and James C. Haggarty
1979 Trampling as an agency in the formation of edge damage: an experiment in lithic technology. *Northwest Anthropological Research Notes* 13 (2): 208-214.

Foix, Louis M. and Ronna J. Bradley
1985 Rhyolite: studies in use-wear analysis. In *Views of the Jornada Mogollon,* edited by C.M. Beck, pp. 112-120. Eastern New Mexico University Contributions to Anthropology, Vol. 12.

Gendel, Peter A., and L. Pirnay
1982 Microwear analysis of experimental stone tools: further test results. *Studia Praehistorica Belgica* 2: 251-265.

Gould, Richard A., Dorothy A. Koster and Ann H. L. Sontz
1971 The lithic assemblage of the Western Desert Aborigines of Australia. *American Antiquity* 36 (2): 149-169.

Grace, R., I.D.G. Graham and M.H. Newcomer
1985 The quantification of microwear polishes. *World Archaeology* 17 (1): 112-120.

Greaves, Sheila
1982 *Upon the Point: A Preliminary Investigation of Ethnicity as a Source of Metric Variation in. Lithic Projectile Points.* National Museum of Man, Mercury Series, Archaeological Survey of Canada, Paper 109.

Greiser, Sally T., and Payson D. Sheets
1979 Raw materials as a functional variable in use-wear studies. In *Lithic Use-Wear Analysis*, edited by B. Hayden, pp. 289-296. Academic Press, New York.

Hair, Joseph F., R.E. Anderson, R.L. Tatham and B.J. Grablowsky
1979 *Multivariate Data Analysis With Readings.* Petroleum Publishing Company, Tulsa, Oklahoma.

Hay, Conran A.
1977 Use-scratch morphology: a functionally significant aspect of edge damage on obsidian tools. *Journal of Field Archaeology* 4: 491-494.

Hayden, Brian
1979a (Ed.) *Lithic Use-wear Analysis*. Academic Press, New York.

1979b Snap, shatter, and superfractures: use-wear of stone skin scrapers. In *Lithic Use-Wear Analysis*, edited by B. Hayden, pp. 207-229. Academic Press, New York.

1986 Use and misuse: the analysis of endscrapers. *Lithic Technology* 15 (2): 65-70.

Hester, Thomas R. and W. I. Follett
1976 Yurok fish knives: a study of wear patterns and adhering salmon scales. In *Experiment and Function: Four California Studies,* pp. 2-23. Contributions of the University of California Archaeological Research Facility, No. 33. University of California, Berkeley.

Hester, Thomas R., Lee Spencer, Colin Busby and James Bard
1976 Butchering a deer with obsidian tools. In *Experiment and Function: Four Californiia Studies*, pp. 45-56. Contributions of the University of California Archaeological Research Facility, No. 33. Unversity of California, Berkeley.

Holley, Gerald A. and Terry A. Del Bene
1981 An evaluation of Keeley's microwear approach. *Journal of Archaeological Science* 8: 337-352.

Kamminga, Johan
1978 *Journey into the Microcosms: A Functional Analysis of Certain Classes of Prehistoric Australian Tools.* Ph.D. thesis, University of Sydney. Sydney.

1979 The nature of use-polish and abrasive smoothing on stone tools. In *Lithic Use-Wear Analysis*, edited by B. Hayden, pp. 143-157. Academic Press, New York.

Keeley, Lawrence H.
1974 Technique and methodology in microwear studies: a critical review. *World Archaeology* 5: 323-336.

1977 The functions of Palaeolithic flint tools. *Scientific American* 237: 108-126.

1980 *Experimental Determination of Stone Tool Uses: A Microwear Analysis.* University of Chicago Press, Chicago.

1981 Reply to Holley and Del Bene. *Journal of Archaeological Science* 8: 348-352.

Keeley, Lawrence H. and Mark H. Newcomer
1977 Microwear analysis of experimental flint tools: a test case. *Journal of Archaeological Science* 4: 29-64.

Keeley, Lawrence H. and Nicholas Toth
1981 Microwear polishes on early stone tools from Koobi Fora, Kenya. *Nature* 293: 464-465.

Kennedy, Dorothy I.D. and Randy Bouchard
1978 Fraser River Lillooet: An Ethnographic Summary. In *Reports of the Lillooet Archaeological Project, No. 1, Introduction and Setting*, edited by A.H. Stryd and S. Lawhead, pp. 22-53. National Museum of Man, Mercury Series, Archaeological Survey of Canada, Paper No. 73.

Knuttson, K. and R. Hope
1984 The application of acetate peels in lithic usewear analysis. *Archaeometry* 26 (1): 49-61.

Lawn, B.R. and D.B. Marshall
1979 Mechanisms of microcontact fracture in brittle solids. In *Lithic Use-Wear Analysis*, edited by B. Hayden, pp. 63-82. Academic Press, New York.

Lawrence, Robert A.
1979 Experimental evidence for the significance of attributes used in edge-damage analysis. In *Lithic Use-Wear Analysis*, edited by B. Hayden, pp. 113-121. Academic Press, New York.

Leach, Chris
1979 *Introduction to Statistics: A Nonparametric Approach for the Social Sciences*. John Wiley and Sons, New York.

Levitt, Jeff
1976 *The Traceological Method: An Introduction and Case Study Analysis of Some Late Prehistoric Hide-Scraping Tools from the Northwest Plains*. M.A. thesis, Department of Archaeology, University of Calgary.

Lewenstein, Suzanne
1981 Mesoamerican obsidian blades: an experimental approach to function. *Journal of Field Archaeology* 8 (2): 175-188.

1984 *Stone Tool Use at Cerros, a Late Preclassic Mayan Site on the North Coast of Belize*. Ph.D. thesis, Department of Anthropology, Arizona State University.

Loy, Thomas H.
1983 Prehistoric blood residues: detection on tool surfaces and identification of species of origin. *Science* 220: 1269-1271.

Lynott, Mark J.
1975 Explanation of microwear patterns on gravers. *Plains Anthropology* 20 (68): 121-128.

Magne, Martin and David Pokotylo
1981 A pilot study in bifacial lithic reduction sequences. *Lithic Technology* 10 (2-3): 34-47.

McConaughy, Mark A.
1979 *Formal and Functional Analyses of the Chipped Stone Tools from Bab Edh-Dhra, Jordan*. Ph.D. thesis, Department of Anthropology, University of Pittsburgh.

Meeks, N. D., G. de G. Sieveking, M. S. Tite and J. Cook
1982 Gloss and use-wear traces on flint sickles and similar phenomena. *Journal of Archaeological Science* 9 (4): 317-340.

Montgomery, Pamela
1978 *Stone Artifacts from the Punchaw Lake Site (Area C): A Late Prehistoric Occupation in Central British Columbia*. M.A. thesis, Department of Archaeology, Simon Fraser University, Burnaby.

Moss, Emily H.
1983a A microwear analysis of burins and points from Tell Abu Hureyra, Syria. In *Traces d'Utilisation sur les Outils Neolithiques du Proche Orient*, edited by M. C. Cauvin, pp. 143-161. Travaux de la Maison de l'Orient No. 5. Maison de l'Orient, Lyon.

1983b *The Functional Analysis of Flint Implements: Pincevent and Pont d'Ambon: Two Case Studies from the French Final Paleolithic.* British Archaeological Reports, International Series, 177.

Myers, Jerome L.
1972 *Fundamentals of Experimental Design* (2nd Edition). Allyn and Bacon, Boston.

Nance, Jack D.
1977 Numerical taxonomy studies of microwear on the Los Tapiales artifacts. Appendix in Los Tapiales: a Paleo-Indian campsite in the Guatemalan highlands by Ruth Gruhn and Alan Lyle Byran, pp. 264-273. *Proceedings of the American Philosophical Society* 121 (3): 235-273.

Newcomer, Mark H.
1976 Spontaneous retouch. *Staringia* 3: 62-64.

Newcomer, Mark H. and Lawrence H. Keeley
1979 Testing a method of microwear analysis with experimental flint tools. In *Lithic Use-Wear Analysis*, edited by B. Hayden, pp. 195-205. Academic Press, New York.

Newcomer, Mark H., R. Grace and R. Unger-Hamilton
1986 Investigating microwear polishes with blind tests. *Journal of Archaeological Science* 13: 203-217.

Nissen, Karen and Margaret Dittemore
1974 Ethnographic data and wear pattern analysis: a study of socketed Eskimo scrapers. *Tebiwa* 17: 67-88.

Norusis, Marija J.
1985 SPSS-X Advanced Statistics Guide. McGraw-Hill, New York.

Odell, George H.
1975 Micro-wear in perspective: a sympathetic response to Lawrence H. Keeley. *World Archaeology* 7: 226-240.

1977 *The Application of Micro-Wear Analysis to the Lithic Component of an Entire Prehistoric Settlement: Methods, Problems and Functional Reconstructions.* Ph.D. thesis, Department of Anthropology, Harvard University.

1980a Toward a more behavioural approach to archaeological lithic concentrations. *American Antiquity* 45: 404-431.

1980b Butchering with stone tools: some experimental results. *Lithic Technology* 9: 39-48.

1981 The mechanics of use-breakage of stone tools: some testable hypotheses. *Journal of Field Archaeology* 8: 197-209.

1982 Emerging directions in the analysis of prehistoric stone tool use. *Reviews in Anthropology* 9: 17-33.

Odell, George H., and Frieda Odell-Vereecken
1980 Verifying the reliability of lithic use-wear assessments by "blind tests": the low-power approach. *Journal of Field Archaeology* 7: 87-120.

Paull, William A.
1984 Organic residue analysis of stone and bone tools. In *Athapaskan and Earlier Archaeology at Big Eagle Lake, British Columbia*, edited by M.P.R. Magne and R.G. Matson, pp. 239-266. Report to the Social Sciences and Humanities Research Council of Canada, Ottawa.

Plisson, Hugues
1983 An application of casting techniques for observing and recording of microwear. *Lithic Technology* 12 (1): 17-21.

Pokotylo, David L. and Patricia Beirne
1978 *Preliminary Inventory, Assessment and Impact Evaluation of Cultural Resources in the Upper Hat Creek Valley, B.C.* Report on file, Heritage Conservation Branch, Victoria.

Price-Beggerly, Patricia
1976 Edge damage on experimentally used scrapers of Hawaiian basalt. *Newsletter of Lithic Technology* 5 (3): 22-24.

Ray, Verne F.
1939 *Cultural Relations in the Plateau of Northwestern America.* Publications of the Frederick Webb Hodge Anniversary Publication Fund 3.

1942 Culture Element Distribution: XXII Plateau. *Anthropological Records* 8: 99-262.

Richards, Thomas H.
1978 *Excavations at EeRl 171.* Report on file, Heritage Conservation Branch, Victoria.

1984 Searching high and low: a review and comparison of microwear analytical methodologies. *The Western Canadian Anthropologist* 1: 18-25.

1987 *Microwear Patterns on Experimental Cache Creek Basalt Tools.* M.A. thesis, Department of Anthropology and Archaeology, University of Saskatchewan, Saskatoon.

Richards, Thomas H. and Mike K. Rousseau
1982 *Archaeological Investigations on Kamloops Indian Reserve No. 1, Kamloops, British Columbia.* Report on file, Heritage Conservation Branch, Victoria.

1983 *The Later Prehistory of the Canadian Plateau.* Paper presented at the XIth International Congress of Anthropological and Ethnological Sciences, Vancouver.

1987 *Late Prehistoric Cultural Horizons on the Canadian Plateau.* Archaeology Press, Simon Fraser University, Burnaby.

Roberts, Michael E.
1980 A preliminary functional analysis of stone tools. *Bulletin of the Massachusetts Archaeological Society* 41 (2): 46-53.

Rousseau, Mike K. and Thomas H. Richards
1985 A culture-historical sequence for the South Thompson River-Western Shuswap Lakes region of British Columbia: The last 4000 years. *Northwest Anthropological Research Notes* 19 (1): 1-32.

Rousseau, Mike K. and Thomas H. Richards
1988 The Oregon Jack Creek site (EdRi-6): a Lehman phase site in the Thompson River Valley, British Columbia. *Canadian Journal of Archaeology* 12: 39-63.

Romesburg, Charles
1984 *Cluster Analysis for Researchers.* Lifetime Learning Publications, Belmont, California.

Ryder, June M.
1976 Terrain inventory and quaternary geology, Ashcroft, British Columbia. *Geological Survey of Canada,* Paper 74-49.

Sabo, Deborah Rowland
1982 Preliminary use-wear analysis of a Michigan chert. *The Michigan Archaeologist* 28 (3):55-72.

Sanger, David
1968a The Chase burial site in British Columbia. *National Museum of Canada Bulletin* 224: 86-185.

1968b Prepared Core and Blade Traditions in the Pacific Northwest. *Arctic Anthropology* 5 (1): 92-120.

1970 *The Archaeology of the Lochnore-Nesikep Locality, British Columbia.* Syesis 3, Supplement 1. British Columbia Provincial Museum, Victoria.

Schutt, Jeanne Ann
1982 *A Comparative Analysis of Wear Patterns on Experimental Lithic Flake Tools: the Re-Examination of Current Concepts in Tool Utilization.* M.A. thesis, Department of Anthropology, University of New Mexico.

Semenov, Sergei A.
1964 *Prehistoric Technology.* Translated by M.W. Thompson. Cory, Adams and MacKay, London.

Semenov, Sergei A. and V.A. Shchelinskii
1971 Micrometrical study of the traces of work on Palaeolithic tools. *Soviet Archaeology* 1971 (1): 19-30.

Shafer, Harry J., and Richard G. Holloway
1979 Organic residue analysis in determining stone tool function. In *Lithic Use-Wear Analysis*, edited by B. Hayden, pp. 385-400. Academic Press, New York.

Shiner, Joel L. and John J Porter
1974 Scanning electron microscopy and prehistoric stone tools. *The Texas Journal of Science* 26: 284-286.

Siegel, Peter E.
1984 Functional variability within an assemblage of endscrapers. *Lithic Technology* 13 (2): 35-51.

1986 More on functional variability within an assemblage of endscrapers: a reply to Hayden and Bamforth. *Lithic Technology* 15 (2): 71-77.

Smith, Harlan
1899 The Archaeology of Lytton, British Columbia. *American Museum of Natural History, Memoirs* 1 (3), New York.

Sneath, Peter H.A. and Robert R. Sokal
1973 *Numerical Taxonomy.* W.H. Freeman, San Francisco.

Stafford, Barbara Domeier
1977 Burin manufacture and utilization: an experimental study. *Journal of Field Archaeology* 4: 235-246.

Stafford, Russell C. and Barbara D. Stafford
1983 The functional hypothesis: a formal approach to use-wear experiments and settlement-subsistence. *Journal of Anthropological Research* 39 (4): 351-375.

Stryd, Arnoud H.
1972 Housepit archaeology at Lillooet, British Columbia: the 1970 field season. *B.C. Studies* 14: 17-46.

1973a *The Later Prehistory of the Lillooet Area, British Columbia.* Ph.D. thesis, University of Calgary.

1973b *Culture History of the Canadian Plateau*. Paper presented at the 6th Annual Meeting of the Canadian Archaeological Association, Burnaby.

1974 *Lillooet Archaeological Project: 1974 Field Season.* Cariboo College Papers in Archaeology 1, Kamloops.

1981 *The 1980 Investigation of the Monte Creek Archaeological Site (EdQx 15).* Report on file, Heritage Conservation Branch, Victoria.

Stryd, Arnoud H. and Stephen Lawhead
1983 *Bethlehem Copper Corporation Lake Zone Development Heritage Mitigation Study.* Report on file, Heritage Conservation Branch, Victoria.

Sussman, Carole
1984 Preliminary results of microwear analysis on experimental quartz tools. *Current Research* 1: 37-39.

1985 Microwear on quartz: fact or fiction? *World Archaeology* 17 (1): 101-111.

Teit, James A.
1900 The Thompson Indians. *American Museum of Natural History, Memoirs* 2 (4): 163-392, New York.

1906 The Lillooet Indians. *American Museum of Natural History, Memoirs* 2 (5): 195-292, New York.

1909 The Shuswap. *American Museum of Natural History, Memoirs* 2 (7): 447-758, New York.

Thomas, David Hurst
1976 *Figuring Anthropology: First Principles of Probability and Statistics.* Holt, Rinehart and Winston, New York.

Tipper, H.W.
1978 Taseko Lakes map-area, British Columbia. *Geological Survey of Canada, Map* 534.

Toll, H. Wolcott
1978 Quartzite qua lithic material in archaeology: qualities and quandaries with special reference to use-wear. *Plains Anthropologist* 23 (79): 47-68.

Tomenchuk, John
1979 The calculation of edge loss on chipped-stone tools resulting from use. In *Lithic Use-Wear Analysis,* edited by B. Hayden, pp. 123-132. Academic Press, New York.

1983 Predicting the past: examples from the use-wear study of selected chipped stone tools from two Epi-Paleolithic occupations in Israel. In *Traces d'Utilisation sur les Outils Neolithiques du Proche Orient,* edited by M. C. Cauvin, pp. 57-76. Travaux de la Maison de l'Orient No. 5. Maison de l'Orient, Lyon.

1984 *A Functional Study of Selected Levantine Epipaleolithic Assemblages.* Ph.D. thesis, Department of Anthropology, University of Toronto.

Tringham, Ruth, G. Cooper, G. Odell, B. Voytek, and A Whitman
1974 Experimentation in the formation of edge damage: a new approach to lithic analysis. *Journal of Field Archaeology* 1: 171-196.

Tsirk, Ari
1979 Regarding fracture initiations. In *Lithic Use-Wear Analysis*, edited by B. Hayden, pp. 83-96. Academic Press, New York.

Turner, Nancy J.
1979 *Plants in British Columbia Indian Technology*. British Columbia Provincial Museum Handbook 38.

Unger-Hamilton, Romana
1983 An investigation into the variables affecting the development and the appearance of plant polish on flint blades. In *Traces d'Utilisation sur les Outils Neolithiques du Proche Orient,* edited by M. C. Cauvin, pp. 243-250. Travaux de la Maison de L'Orient No. 5. Maison de l'Orient, Lyon.

1984 The formation of use-wear polish on flint: beyond the "deposit verus abrasion" controversy. *Journal of Archaeological Science* 11: 91-98.

1985 Microscopic striations on flint sickle-blades as an indication of plant cultivation: preliminary results. *World Archaeology* 17 (1): 121-126.

Vaughan, P.C.
1981 Microwear analysis of experimental flint and obsidian tools. *Staringia* 6: 90-91.

1985 *Use-Wear Analysis of Flaked Stone Tools.* University of Arizona Press, Tuscon.

Von Krogh, G. Henning
1978 *Archaeological Investigations in the Spences Bridge Locality, British Columbia.* Report on file, Heritage Conservation Branch, Victoria.

Walde, Dale Allen
1985 *On Assigning Gender to Post-Cranial Bison Bones.* M.A. thesis, Department of Anthropology, University of Victoria.

Walker, Ernest G.
1980 *The Gowen Site: an Early Archaic Site on the Northern Plains.* Ph.D. thesis, Department of Anthropology, University of Texas at Austin.

White, Randall
1982 The manipulation of burins in incision and notation. *Canadian Journal of Anthropology* 2 (2): 129-136.

Whitlam, Robert G.
1980 *Archaeological Investigations at Cache Creek (EeRh 3).* Occasional Papers of the Heritage Conservation Branch 5, Victoria.

Wilson, Robert L.
1980 Archaeological Investigations Near Kamloops. In *The Archaeology of Kamloops,* by R. Wilson and C. Carlson, pp. 1-83. Department of Archaeology, Simon Fraser University, Publication 7, Burnaby.

Wolff, Diane D., and Michael L. Parsons
1983 *Pattern Recognition Approach to Data Interpretation.* Plenum Press, New York.

Wyant, Jeffrey R. and Frank E. Bayham
1976 An experimental study of lithic use wear. In *Desert Resources and Hohokam Subsistence: The Conoco Florence Project,* assembled by William H. Doelle, pp. 231-243. Arizona State Museum Archaeological Series 103.

Wyatt, David J.
1972 *The Indian History of the Nicola Valley, British Columbia.* Ph.D. thesis, Department of Anthropology, Brown University.

Yerkes, Richard W.
1983 Microwear, microdrills, and Mississippian craft specialization. *American Antiquity* 48: 499-518.

APPENDIX A

DETAILED TABULATIONS OF MICROWEAR OBSERVATIONS

TABLES 28-30

Table 28. Microflake scar distribution on experimental tools observed with a stereomicroscope at 40 x. For each tool the raw number of scars of each size and type are listed, with ventral scars on top and dorsal scars on the line immediately below. The total number of scars on each tool can be found in the far right column.

TOOL	STEP				HINGE				FEATHER				SNAP				TOTAL
NUMBER	S	M	L	V	S	M	L	V	S	M	L	V	S	M	L	V	SCARS
PLANTS																	
12-9									5				2	2			15
									1				3	2			
12-12									5								5
134-11									7				11	1	1		35
					5				6	2	1		1				
43-2					1				20	3							45
			2				1		15	2	1						
161-11	13	5							14	1			2	3	1		70
	15				1				12				1	1	1		
HIDE																	
1-3																	1
	1																
12-13																	15
	7	2	3	1					1		1						
17-14		1							4								77
	56	6	5						2	1	2						
22-7									1								71
	41	4			3	1			20	1							
23-3																	46
	25	4	3		10	2			2								
27-16	9	1	1		11				13				3				78
	14	3			6				13	2	1			1			
39-1																	25
	19	2				1		1			2						

continued

Table 28. (continued)

TOOL NUMBER	STEP S	STEP M	STEP L	STEP V	HINGE S	HINGE M	HINGE L	HINGE V	FEATHER S	FEATHER M	FEATHER L	FEATHER V	SNAP S	SNAP M	SNAP L	SNAP V	TOTAL SCARS
HIDE (continued)																	
44-4											1						66
	15	2							46				1	1			
3-3										2	2						12
	1	1							3	2	1						
3-4	1								6	1							12
									3	1							
22-12	6				1				9				1				26
	1	2	2	2	1								1				
27-17	10				3				10								28
	1		1						3								
31-2			1														3
							2										
44-6																	41
	3	2	1		1				34								
46-4									7								33
	7			3					11	1	4						
80-6	1				7				5				4				38
					2				3				13	2	1		
80-4*																	
80-5*																	
80-7	2	1			17				7				2				46
	1				13				1				2				
80-8					1				1								22
					3				2				15				
12-17	4				2				1								43
	5	1			6				17				3	1	3		
134-9	21	3	2	1	2	1	1		21	4		1	1	2	1		123
	39		2		1	3			15	2							

continued

Table 20. (continued)

TOOL NUMBER	STEP S	STEP M	STEP L	STEP V	HINGE S	HINGE M	HINGE L	HINGE V	FEATHER S	FEATHER M	FEATHER L	FEATHER V	SNAP S	SNAP M	SNAP L	SNAP V	TOTAL SCARS
BONE																	
21-6	14	1	5		14	2	3		7	3				1	1	1	74
	12				5	1			1					1	1	1	
26-8	2				4	3			2	3	1				1	1	19
					1					1							
26-9	2				4	1					2				3	1	22
	1	2											1	3	2		
46-2	18	2	2		23	2	1		1		1		2		2	1	94
	14				15	2			1		2		1	1	2	1	
11-6					1									2			36
	16	2	1		2	2			4	1			1		1	3	
18-19	25	2		1			1		11							1	55
	6	2		1					3	1						1	
22-20	21	2	1		20	1		1	12	1			1				98
	8				15				14	1							
17-13	1	1							2								61
	34	5	10		4				4								
63-1	6				7	1			2	1			1				68
	4		1		32	1	3	2	3				2	1	1		
ANTLER																	
43-3	9	5							17	2	11					1	83
	11		1						16	4	5					1	
134-22	24	15	1		2		1		3	2	12					3	101
	18	5	4		3	2	1		4		1						
141-7	13	1			4				9	3	2	1	5	1	1		68
	13				4				4		3		2	1		1	
11-2	1		1						17	2	1			2			37
	4	1			1				6	1							

continued

Table 28. (continued)

TOOL NUMBER	STEP S	STEP M	STEP L	STEP V	HINGE S	HINGE M	HINGE L	HINGE V	FEATHER S	FEATHER M	FEATHER L	FEATHER V	SNAP S	SNAP M	SNAP L	SNAP V	TOTAL SCARS
ANTLER (continued)																	
11-12	7	2			3				4	2			3		2	1	53
	3	2			3	1			7				4	4	4	1	
17-11	8	1	1		2	1	2		3		1		1	2			38
	3	2			2				2	1	1		1	1	1	2	
27-3	4				2	1	1		5				5		1		42
	9					2			3	2			4	1	2		
36-7	19	2	1		9	1			17	1			6	1	2		86
	2	1	1		7	1			11		1		2		1		
18-17	3	3			3				8	1			5	3	1		61
	7	2			1				5	1			9	8	1		
141-12	4				9				23		2			3	6		90
	16	3			18				4					1		1	
22-11	5	1	1						10	3					3		79
	5	14	3	10	3	1		1	19								
138-13																	58
	31	9	15	2							1						
141-14	2				2												72
	24	3			20	1			13	2	2		2	1			
161-18	14	6	2	1	2			1	8	1	1						84
	12	10	1	2	8	3	4	4	2	1	1						
161-19	12	2	1		8	1	2		13	1							143
	33	4	7	1	31	3	1	1	16		5		1				
FISH																	
12-10	7	2			2				20	1				1			41
	3								1				4				
12-14																	0
134-16	9					1			5								62
	22	1			1				22		1						
20-7																	0

continued

Table 28. (continued)

TOOL	STEP				HINGE				FEATHER				SNAP				TOTAL
NUMBER	S	M	L	V	S	M	L	V	S	M	L	V	S	M	L	V	SCARS
FISH (continued)																	
46-3									16								31
	4								9				2				
138-12	22	1			3				44				4				110
	24					1			8				2	1			
134-17	6								27				2				45
					1				9								
DEER																	
12-3	39				7				32				9				136
	16				2				19				9	2	1		
12-6	3				1				11				25	1			63
	1				1				12				8				
43-4	5				3				48								83
	7								20								
80-1	3				5				23				8				64
	7				1				12				5				
80-12	9				4				15				9	1	1		92
	9				8	2			14				15	2	3		
80-13	4								7				3	2			30
	1								13								
1-4																	0
11-9									2								2
22-17	4												2				51
	8				4				14	2			17				
27-5	6								1								15
	6								2								
35-3									2								2
36-8	11								12								30
	5								2								

continued

Table 28. (continued)

TOOL NUMBER	STEP S	M	L	V	HINGE S	M	L	V	FEATHER S	M	L	V	SNAP S	M	L	V	TOTAL SCARS
DEER (continued)																	
45-7	60				1				27				3				141
	31	2	1		2				13				1				
12-15	9				1				16	2							55
	14								9				4				
22-10	5								4				2				33
	11				4				7								
22-15	8		2		6	2			13	2			2				59
	9	2			3	2			5	1			1	1			
45-6	16	1			1				8				4				49
	6		1		1	1			8				1	1			
12-11	19	2	1		14	1	1		27	4			2				138
	27	4		1	8	3	1		17	2	3		1				
80-3	13	3			9	1			13	1			3				125
	32	9	4	3	6		1		16	4	2		4	1			
80-14*																	
44-8					3				7								62
	24	5	1		7	4			5	4	2						
44-11	9				8	2	1		7	3	2		1				81
	24	2	1		7	4	5		4		1						
2-1	6								3								18
	2								3				3	1			
2-3					2				2				3				25
					3				15								
11-3	1								1				2				20
	3				1				10				2				
12-1	18				11				15	2	1		2				82
	12				7				13					1			

continued

Table 28. (continued)

TOOL NUMBER	STEP S	STEP M	STEP L	STEP V	HINGE S	HINGE M	HINGE L	HINGE V	FEATHER S	FEATHER M	FEATHER L	FEATHER V	SNAP S	SNAP M	SNAP L	SNAP V	TOTAL SCARS
WOOD																	
27-11																	52
	1				10	1			31	5	2		2				
36-5	8				7				9								95
	16				19	4	1		31								
44-10																	22
					8				14								
141-6	7								7								190
	101	2	1						68	3				1			
141-11	10				4				22								86
	7				23	1			19								
161-20					2	3						1					99
	30	2	2		30				24	4		1					
26-10									2								28
	22				2				3	1							
27-15									2								30
	22				2				3	1							
134-23	3								2								43
	28				1				9								
138-6	1								3								65
	13	5							42	1							
138-16												2					87
	31	2	3						46	2		1					
141-4									3	1							29
	8				4				13								
141-17	11																131
	94	6							18	2							
27-10		1	1						3		1			1		1	15
									2				3	1		1	
134-18	6				2				15	2			4	1	1		73
	16				3				19				1	1	2		

continued

Table 28. (continued)

TOOL NUMBER	STEP S	STEP M	STEP L	STEP V	HINGE S	HINGE M	HINGE L	HINGE V	FEATHER S	FEATHER M	FEATHER L	FEATHER V	SNAP S	SNAP M	SNAP L	SNAP V	TOTAL SCARS
WOOD (continued)																	
63-2	16	3	1		1				9	3	1		2		4	1	83
	19	2			2				14	2				1		2	
82-3	1							2	2	1							29
	5			1					15	1			1				
27-7	2	1							9				9	2			50
	8	1							13					3	1	1	
27-14	5				15	1			16				3	4			91
	10				14	2		1	14	4			1		1		
12-5					1				8				3		3		60
	20	4							15				5	1			
17-12	1			1				1	11							1	56
	1	2	1		2	1			28	3			3				
80-9	3				1				3								50
	23	2							10	3	1		1	2	1		
80-10	4								2				1	1			43
	8				2				19	1			4	1			
80-11**																	
23-4																	0
27-8	3				4				22				6				73
	2				14				20				2				
44-7	12				3				7				13	1	1		126
	15				2				70				2				
44-9	17								9				2				90
	38	2			1				21								
161-15	39	4	1	1	14	4	2	2	11	1	3		1				153
	21	1	2	1	24	6	4	4	7				1				

* = deliberately retouched prior to use
**= extensive pre-use damage

Table 29. Micropolish characteristics observed on experimental basalt tools with an incident light microscope at 200-400 x. A key to the abbreviations used is at the end of the table.

TOOL NUMBER		GRAINY D	GRAINY D-B	GRAINY B	SMOOTH B-P	SMOOTH B	TRACE ONLY
PLANTS							
12-9	v	-	-	-	E	-	-
	e	-	-	-	E	-	-
	d	-	-	-	E	-	-
12-12	v	-	-	-	S	-	-
	e	-	-	-	E	-	-
	d	-	-	-	E	-	-
134-11	v	-	-	-	E	-	-
	e	-	-	-	E	-	-
	d	-	-	-	E	-	-
43-2	v	-	-	-	T	-	-
	e	-	-	-	S	-	-
	d	-	-	-	T	-	-
161-11	v	-	-	-	E	-	-
	e	-	-	-	E	-	-
	d	-	-	-	E	-	
HIDE							
1-3	v	-	-	-	-	-	+
	e	-	-	-	-	-	+
	d	-	-	-	-	-	+
12-13	v	E	-	-	-	-	-
	e	E	-	-	-	-	-
	d	E	-	-	-	-	-
17-14	v	E	-	-	-	-	-
	e	E	-	-	-	T	-
	d	E	-	-	-	-	-

continued

Table 29. (continued)

TOOL NUMBER		GRAINY D	D-B	B	SMOOTH B-P	B	TRACE ONLY
HIDE (continued)							
22-7	v	S	-	-	-	-	-
	e	E	-	-	-	-	-
	d	S	-	-	-	-	-
23-3	v	-	-	-	-	-	+
	e	-	-	-	-	-	+
	d	-	-	-	-	-	+
27-16	v	S	-	-	-	-	-
	e	E	-	-	-	-	-
	d	E	-	-	-	-	-
39-1	v	E	-	-	-	-	-
	e	E	-	-	-	-	-
	d	E	-	-	-	-	-
44-4	v	S	-	-	-	-	-
	e	E	-	-	-	-	-
	d	S	-	-	-	-	-
3-3	v	-	-	-	-	-	-
	e	-	-	-	-	-	-
	d	-	-	-	-	-	-
3-4	v	-	T	-	-	-	-
	e	-	E	-	-	-	-
	d	-	T	-	-	-	-
22-12	v	E	-	-	-	-	-
	e	E	-	-	-	-	-
	d	E	-	-	-	-	-
27-17	v	E	-	-	-	-	-
	e	E	-	-	-	-	-
	d	E	-	-	-	-	-
31-2	v	-	-	-	-	-	-
	e	-	-	-	-	-	-
	d	-	-	-	-	-	-

continued

Table 29. (continued)

TOOL NUMBER		GRAINY D	D-B	B	SMOOTH B-P	B	TRACE ONLY
HIDE (continued)							
44-6	**v**	**E**	-	-	-	-	-
	e	**E**	-	-	-	-	-
	d	**S**	-	-	-	-	-
46-4	**v**	**E**	-	-	-	-	-
	e	**E**	-	-	-	-	-
	d	**E**	-	-	-	-	-
80-6	**v**	-	-	-	-	-	-
	e	-	-	-	-	-	-
	d	-	-	-	-	-	-
80-4	**v**	**S**	-	-	-	-	-
	e	**E**	-	-	-	-	-
	d	**E**	-	-	-	-	-
80-5	**v**	**S**	-	-	-	-	-
	e	**S**	-	-	-	-	-
	d	**T**	-	-	-	-	-
80-7	**v**	-	-	-	-	-	-
	e	**S**	-	-	-	-	-
	d	-	-	-	-	-	-
80-8	**v**	**S**	-	-	-	-	-
	e	**S**	-	-	-	-	-
	d	-	-	-	-	-	-
12-17	**v**	**S**	-	-	-	-	-
	e	**E**	-	-	-	-	-
	d	**S**	-	-	-	-	-
134-9	**v**	**S**	-	-	-	-	-
	e	**E**	-	-	-	-	-
	d	**E**	-	-	-	-	-

continued

Table 29. (continued)

TOOL NUMBER		GRAINY D	GRAINY D-B	GRAINY B	SMOOTH B-P	SMOOTH B	TRACE ONLY
BONE							
21-6	v	-	-	S	-	-	-
	e	-	-	S	-	-	-
	d	-	-	-	-	-	-
26-8	v	-	-	S	-	S	-
	e	S	-	S	-	S	-
	d	-	-	S	-	S	-
26-9	v	-	-	S	-	-	-
	e	-	-	S	-	-	-
	d	-	-	S	-	-	-
46-2	v	-	-	S	-	-	-
	e	-	-	S	-	-	-
	d	-	-	S	-	-	-
11-6	v	S	-	-	-	S	-
	e	S	-	-	-	S	-
	d	S	-	-	-	S	-
18-19	v	T	-	-	-	-	-
	e	-	T	-	-	-	-
	d	T	-	-	-	-	-
22-20	v	T	-	-	-	T	-
	e	S	-	-	-	S	-
	d	S	-	-	-	S	-
17-13	v	-	-	-	-	-	-
	e	-	-	E	-	-	-
	d	-	-	S	-	-	-
63-1	v	-	-	-	-	-	-
	e	S	-	-	-	S	-
	d	T	-	-	-	T	-
138-19	v	-	E	-	E	-	-
	e	-	E	-	E	-	-
	d	-	E	-	E	-	-

continued

Table 29. (continued)

TOOL NUMBER		GRAINY D	GRAINY D-B	GRAINY B	SMOOTH B-P	SMOOTH B	TRACE ONLY
ANTLER							
43-3	v	-	-	-	-	-	-
	e	-	T	-	-	-	-
	d	-	T	-	-	-	-
134-22	v	-	S	-	-	-	-
	e	-	S	-	-	-	-
	d	-	S	-	-	-	-
141-7	v	-	-	T	-	-	-
	e	-	-	T	-	-	-
	d	-	-	T	-	-	-
11-2	v	-	-	-	-	-	-
	e	-	-	-	-	-	-
	d	-	-	-	-	-	-
11-12	v	-	-	T	-	-	-
	e	-	-	-	-	-	-
	d	-	-	T	-	-	-
17-11	v	-	-	T	-	-	-
	e	-	-	S	-	-	-
	d	-	-	S	-	-	-
27-3	v	-	S	-	-	-	-
	e	-	S	-	-	-	-
	d	-	S	-	-	-	-
36-7	v	-	S	-	-	-	-
	e	-	S	-	-	-	-
	d	-	S	-	-	-	-
18-17	v	-	-	-	-	-	-
	e	-	-	S	-	-	-
	d	-	-	S	-	-	-

continued

Table 29. (continued)

TOOL NUMBER		GRAINY D	D-B	B	SMOOTH B-P	B	TRACE ONLY
ANTLER (continued)							
141-12	v	-	S	-	-	-	-
	e	-	S	-	-	-	-
	d	-	S	-	-	-	-
22-11	v	-	-	T	-	-	-
	e	-	-	S	-	-	-
	d	-	-	S	-	-	-
138-13	v	-	-	T	-	-	-
	e	-	-	S	-	-	-
	d	-	-	S	-	-	-
141-14	v	-	-	T	-	-	-
	e	-	-	S	-	-	-
	d	-	-	S	-	-	-
161-18	v	-	-	T	-	-	-
	e	-	-	S	-	-	-
	d	-	-	S	-	-	-
161-19	v	-	T	-	-	-	-
	e	-	E	-	-	-	-
	d	-	S	-	-	-	-
134-6	v	-	-	E	S	-	-
	e	-	-	E	S	-	-
	d	-	-	E	S	-	-
141-20	v	-	-	E	-	S	-
	e	-	-	E	-	S	-
	d	-	-	E	-	S	-
FISH							
12-10	v	T	-	-	-	-	-
	e	S	S	-	-	-	-
	d	T	-	-	-	-	-
12-14	v	T	-	-	-	-	-
	e	T	-	-	-	-	-
	d	T	-	-	-	-	-

continued

Table 29. (continued)

TOOL NUMBER		GRAINY D	GRAINY D-B	GRAINY B	SMOOTH B-P	SMOOTH B	TRACE ONLY
FISH (continued)							
134-16	v	S	-	-	-	-	-
	e	S	S	-	-	T	-
	d	S	-	-	-	T	-
20-7	v	-	-	-	-	-	-
	e	-	-	-	-	-	-
	d	-	-	-	-	-	-
46-3	v	S	-	-	-	-	-
	e	S	S	-	-	-	-
	d	S	-	-	-	-	-
138-12	v	-	-	-	-	-	-
	e	-	-	-	-	-	-
	d	-	-	-	-	-	-
134-17	v	-	-	-	-	-	-
	e	-	-	-	-	T	-
	d	-	-	-	-	-	-
DEER							
12-3	v	E	-	-	-	T	-
	e	-	-	E	-	T	-
	d	S	-	-	-	-	-
12-6	v	S	-	-	-	T	-
	e	S	-	-	-	T	-
	d	S	-	-	-	T	-
43-4	v	-	E	-	-	-	-
	e	-	E	-	-	-	-
	d	-	E	-	-	-	-
80-1	v	T	-	-	-	-	-
	e	T	-	-	-	-	-
	d	S	-	-	-	-	-

continued

Table 29. (continued)

TOOL NUMBER		GRAINY D	D-B	B	SMOOTH B-P	B	TRACE ONLY
DEER (continued)							
80-12	v	E	-	S	-	T	-
	e	-	-	-	-	-	-
	d	S	-	-	-	-	-
80-13	v	-	-	-	-	-	+
	e	-	-	-	-	-	+
	d	-	-	-	-	-	+
1-4	v	E	-	-	-	-	-
	e	E	-	T	-	-	-
	d	E	-	-	-	-	-
11-9	v	-	-	S	-	-	-
	e	-	-	S	-	-	-
	d	-	-	S	-	-	-
22-17	v	T	-	T	-	-	-
	e	-	-	T	-	-	-
	d	T	-	T	-	-	-
27-5	v	-	-	S	-	S	-
	e	T	-	T	-	-	-
	d	T	-	-	-	-	-
35-3	v	-	-	-	-	-	+
	e	-	-	-	-	-	+
	d	-	-	-	-	-	+
36-8	v	-	-	-	-	-	-
	e	-	T	-	-	-	-
	d	-	-	-	-	-	-
45-7	v	-	-	E	-	-	-
	e	-	-	E	-	-	-
	d	-	S	-	-	-	-
12-15	v	T	-	-	-	-	-
	e	T	-	-	-	-	-
	d	T	-	-	-	-	-

continued

Table 29. (continued)

TOOL NUMBER		GRAINY D	D-B	B	SMOOTH B-P	B	TRACE ONLY
DEER (continued)							
22-10	v	T	-	-	-	-	-
	e	T	-	-	-	-	-
	d	T	-	-	-	-	-
22-15	v	T	-	-	-	-	-
	e	-	-	-	-	-	-
	d	T	-	-	-	-	-
45-6	v	S	-	S	-	S	-
	e	-	-	E	-	S	-
	d	S	-	S	-	T	-
12-11	v	-	-	E	-	T	-
	e	-	-	E	-	T	-
	d	-	-	E	-	T	-
80-3	v	T	T	-	-	-	-
	e	S	S	-	-	-	-
	d	T	T	-	-	-	-
80-14	v	-	-	-	-	-	-
	e	S	-	-	-	T	-
	d	-	-	-	-	-	-
44-8	v	-	-	-	-	-	-
	e	-	S	-	-	-	-
	d	-	-	-	-	-	-
44-11	v	S	-	-	-	-	-
	e	T	-	-	-	-	-
	d	T	-	-	-	-	-
2-1	v	-	-	-	-	-	-
	e	-	-	-	-	-	-
	d	-	-	-	-	-	-

continued

Table 29. (continued)

TOOL NUMBER		GRAINY D	GRAINY D-B	GRAINY B	SMOOTH B-P	SMOOTH B	TRACE ONLY
DEER (continued)							
2-3	v	-	-	-	-	-	-
	e	-	-	-	-	-	-
	d	-	-	-	-	-	-
11-3	v	-	-	-	-	-	-
	e	-	-	-	-	-	-
	d	-	-	-	-	-	-
12-1	v	T	-	T	-	-	-
	e	T	-	T	-	-	-
	d	T	-	T	-	-	-
WOOD							
27-11	v	-	T	-	-	-	-
	e	-	-	S	-	S	-
	d	-	T	-	-	-	-
36-5	v	-	-	-	-	T	-
	e	-	-	S	-	S	-
	d	S	-	-	E	-	-
44-10	v	-	-	-	-	-	+
	e	-	-	E	-	E	-
	d	-	-	-	-	-	+
141-6	v	S	-	-	-	-	-
	e	-	S	-	E	-	-
	d	E	-	-	S	-	-
141-11	v	-	-	-	-	S	-
	e	-	-	S	-	E	-
	d	-	-	-	-	-	+
161-20	v	-	-	-	-	T	-
	e	S	-	S	-	S	-
	d	S	-	-	E	-	-

continued

Table 29. (continued)

TOOL NUMBER		GRAINY D	D-B	B	SMOOTH B-P	B	TRACE ONLY
WOOD (continued)							
26-10	v	-	-	T	-	T	-
	e	-	-	E	S	-	-
	d	-	-	-	T	-	-
27-15	v	-	-	-	-	-	+
	e	-	-	S	E	-	-
	d	-	-	-	-	-	+
134-23	v	-	-	E	-	E	-
	e	-	-	S	E	-	-
	d	-	-	E	-	E	-
138-6	v	-	-	-	-	S	-
	e	-	-	S	-	S	-
	d	-	-	-	-	T	-
138-16	v	-	-	-	S	-	-
	e	-	-	S	-	-	-
	d	-	-	S	S	-	-
141-4	v	-	-	-	-	-	+
	e	-	-	E	-	S	-
	d	-	-	-	-	-	+
141-17	v	-	-	-	-	-	+
	e	-	-	-	-	-	+
	d	-	-	-	-	-	+
27-10	v	-	-	S	-	S	-
	e	-	-	S	-	-	-
	d	-	-	-	-	-	+
134-18	v	-	S	-	E	-	-
	e	-	T	-	-	T	-
	d	-	S	-	E	-	-
63-2	v	-	-	-	-	-	+
	e	-	-	-	-	-	+
	d	-	-	-	-	-	+

continued

Table 29. (continued)

TOOL NUMBER		GRAINY D	D-B	B	SMOOTH B-P	B	TRACE ONLY
WOOD (continued)							
82-3	v	-	-	-	-	-	+
	e	-	-	-	-	S	-
	d	-	-	-	-	-	+
27-7	v	-	T	-	-	T	-
	e	S	-	S	-	S	-
	d	-	-	-	-	S	-
27-14	v	S	-	-	-	-	-
	e	-	E	-	-	-	-
	d	S	-	-	-	-	-
12-5	v	-	S	-	S	-	-
	e	-	-	-	-	-	-
	d	-	E	-	S	-	-
17-12	v	T	-	-	-	-	-
	e	S	-	S	-	T	-
	d	T	-	-	-	-	-
80-9	v	-	-	-	S	-	-
	e	-	-	S	E	-	-
	d	-	-	-	S	-	-
80-10	v	-	-	-	-	-	+
	e	-	-	-	-	S	-
	d	T	-	-	-	T	-
80-11	v	-	S	-	T	-	-
	e	-	S	-	E	-	-
	d	T	-	-	-	-	-
23-4	v	-	-	-	-	-	+
	e	-	-	-	-	-	+
	d	-	-	-	-	-	+

continued

Table 29. (continued)

TOOL NUMBER		GRAINY D	GRAINY D-B	GRAINY B	SMOOTH B-P	SMOOTH B	TRACE ONLY
WOOD (continued)							
27-8	v	-	-	-	E	-	-
	e	-	-	-	S	-	-
	d	-	-	-	-	-	+
44-7	v	T	-	-	-	S	-
	e	S	-	-	-	S	-
	d	T	-	-	-	S	-
44-9	v	T	-	-	-	T	-
	e	S	-	-	-	S	-
	d	T	-	-	-	T	-
161-15	v	S	-	-	-	S	-
	e	S	-	-	S	-	-
	d	T	-	-	-	S	-

KEY

v = Ventral aspect of used edge.

e = Used edge.

d = Dorsal aspect of used edge.

GRAINY = Grainy-textured polish, which is either "D" dull, "D-B" dull-bright, or "B" bright in intensity.

SMOOTH = Smooth-textured polish, which is either "B-P" bright in intensity and pitted or "B" bright in intensity and unpitted.

TRACE ONLY = Only the slightest trace of polish is present (+) and it is impossible to classify as to type.

E = Polished areas are extensive.

S = Polish is scattered.

T = Only traces of polish are present.

Table 30. Characteristics of striations observed on experimental tools with light microscopes and a scanning electron microscope. A key to abbreviations used follows the table.

TOOL NUMBER		FREQ	ORIENT	SIZE
PLANTS				
12-9	v	C	P,O	S
	e	-	-	-
	d	C	P,O	S
12-12	v	C	P,O,R	S
	e	-	-	-
	d	C	P,O,R	S
134-11	v	R	P	S
	e	-	-	-
	d	R	P	S
43-2	v	-	-	-
	e	-	-	-
	d	-	-	-
161-11	v	-	-	-
	e	-	-	-
	d	R	P,O	S
HIDE				
1-3	v	-	-	-
	e	-	-	-
	d	-	-	-
12-13	v	-	-	-
	e	-	-	-
	d	-	-	-
17-14	v	R	R	S
	e	-	-	-
	d	R	R	S
22-7	v	R	R	S
	e	-	-	-
	d	R	R,O	M

continued

Table 30. (continued)

TOOL NUMBER		FREQ	ORIENT	SIZE
HIDE (continued)				
23-3	v	-	-	-
	e	-	-	-
	d	-	-	-
27-16	v	-	-	-
	e	-	-	-
	d	C	R,O	S-M
39-1	v	-	-	-
	e	-	-	-
	d	R	R	S-M
44-4	v	-	-	-
	e	-	-	-
	d	-	-	-
3-3	v	-	-	-
	e	-	-	-
	d	-	-	-
3-4	v	-	-	-
	e	-	-	-
	d	-	-	-
22-12	v	-	-	-
	e	-	-	-
	d	C	R	S
27-17	v	R	R	S
	e	-	-	-
	d	C	O,R	S-L
31-2	v	-	-	-
	e	-	-	-
	d	-	-	-
44-6	v	-	-	-
	e	R	R	S
	d	-	-	-

continued

Table 30. (continued)

TOOL NUMBER		FREQ	ORIENT	SIZE
HIDE (continued)				
46-4	v	R	O	S
	e	-	-	-
	d	R	O,R	S
80-6	v	-	-	-
	e	-	-	-
	d	-	-	-
80-4	v	-	-	-
	e	-	-	-
	d	-	-	-
80-5	v	-	-	-
	e	-	-	-
	d	-	-	-
80-7	v	-	-	-
	e	-	-	-
	d	-	-	-
80-8	v	-	-	-
	e	-	-	-
	d	-	-	-
12-17	v	C	O,P	S
	e	-	-	-
	d	C	O,P,R	S
134-9	v	C	O,P	S
	e	-	-	-
	d	C	O,P,R	S
BONE				
21-6	v	-	-	-
	e	-	-	-
	d	-	-	-
26-8	v	-	-	-
	e	-	-	-
	d	-	-	-

continued

Table 30. (continued)

TOOL NUMBER		FREQ	ORIENT	SIZE
BONE (continued)				
26-9	v	-	-	-
	e	-	-	-
	d	-	-	-
46-2	v	-	-	-
	e	-	-	-
	d	-	-	-
11-6	v	-	-	-
	e	-	-	-
	d	-	-	-
18-19	v	-	-	-
	e	-	-	-
	d	R	P,O	S
22-20	v	-	-	-
	e	-	-	-
	d	-	-	-
17-13	v	-	-	-
	e	-	-	-
	d	-	-	-
63-1	v	-	-	-
	e	-	-	-
	d	-	-	-
138-19	v	-	-	-
	e	-	-	-
	d	-	-	-
ANTLER				
43-3	v	-	-	-
	e	-	-	-
	d	-	-	-
134-22	v	-	-	-
	e	-	-	-
	d	-	-	-

continued

Table 30. (continued)

TOOL NUMBER		FREQ	ORIENT	SIZE
ANTLER (continued)				
141-7	v	-	-	-
	e	-	-	-
	d	-	-	-
11-2	v	-	-	-
	e	-	-	-
	d	-	-	-
11-12	v	-	-	-
	e	-	-	-
	d	-	-	-
17-11	v	-	-	-
	e	-	-	-
	d	-	-	-
27-3	v	-	-	-
	e	-	-	-
	d	-	-	-
36-7	v	-	-	-
	e	-	-	-
	d	-	-	-
18-17	v	-	-	-
	e	-	-	-
	d	-	-	-
141-12	v	-	-	-
	e	-	-	-
	d	-	-	-
22-11	v	-	-	-
	e	-	-	-
	d	-	-	-
138-13	v	-	-	-
	e	-	-	-
	d	-	-	-
141-14	v	-	-	-
	e	-	-	-
	d	-	-	-

continued

Table 30. (continued)

TOOL NUMBER		FREQ	ORIENT	SIZE
ANTLER (continued)				
161-18	v	-	-	-
	e	C	R	
	d	-	-	-
161-19	v	-	-	-
	e	-	-	-
	d	R	R	S
134-6	v	-	-	-
	e	-	-	-
	d	-	-	-
141-20	v	-	-	-
	e	R	P,O	S
	d	-	-	-
FISH				
12-10	v	-	-	-
	e	-	-	-
	d	-	-	-
12-14	v	-	-	-
	e	-	-	-
	d	-	-	-
134-16	v	-	-	-
	e	-	-	-
	d	-	-	-
20-7	v	-	-	-
	e	-	-	-
	d	-	-	-
46-3	v	-	-	-
	e	-	-	-
	d	-	-	-
138-12	v	R	P	M
	e	-	-	-
	d	-	-	-
134-17	v	-	-	-
	e	-	-	-
	d	-	-	-

continued

Table 30. (continued)

TOOL NUMBER		FREQ	ORIENT	SIZE
DEER				
12-3	v	R	P	S
	e	-	-	-
	d	R	P	S
12-6	v	C	P,O	S
	e	-	-	-
	d	C	P,O	S
43-4	v	-	-	-
	e	-	-	-
	d	-	-	-
80-1	v	-	-	-
	e	-	-	-
	d	-	-	-
80-12	v	-	-	-
	e	-	-	-
	d	-	-	-
80-13	v	-	-	-
	e	-	-	-
	d	-	-	-
1-4	v	-	-	-
	e	-	-	-
	d	-	-	-
11-9	v	-	-	-
	e	-	-	-
	d	-	-	-
22-17	v	-	-	-
	e	-	-	-
	d	-	-	-
27-5	v	-	-	-
	e	-	-	-
	d	-	-	-

continued

Table 30. (continued)

TOOL NUMBER		FREQ	ORIENT	SIZE
DEER (continued)				
35-3	v	-	-	-
	e	-	-	-
	d	-	-	-
36-8	v	-	-	-
	e	-	-	-
	d	-	-	-
45-7	v	C	P,O	S
	e	-	-	-
	d	C	P,O	S
12-15	v	-	-	-
	e	-	-	-
	d	-	-	-
22-10	v	-	-	-
	e	-	-	-
	d	-	-	-
22-15	v	-	-	-
	e	-	-	-
	d	-	-	-
45-6	v	-	-	-
	e	-	-	-
	d	-	-	-
12-11	v	R	P	S-M
	e	-	-	-
	d	R	P,O	S
80-3	v	-	-	-
	e	-	-	-
	d	-	-	-
80-14	v	-	-	-
	e	-	-	-
	d	-	-	-

continued

Table 30. (continued)

TOOL NUMBER		FREQ	ORIENT	SIZE
DEER (continued)				
44-8	v	-	-	-
	e	-	-	-
	d	-	-	-
44-11	v	-	-	-
	e	-	-	-
	d	-	-	-
2-1	v	-	-	-
	e	-	-	-
	d	-	-	-
2-3	v	-	-	-
	e	-	-	-
	d	-	-	-
11-3	v	-	-	-
	e	-	-	-
	d	-	-	-
12-1	v	R	O	S
	e	-	-	-
	d	-	-	-
WOOD				
27-11	v	-	-	-
	e	C	R	S
	d	-	-	-
36-5	v	-	-	-
	e	-	-	-
	d	C	R,O	S
44-10	v	-	-	-
	e	-	-	-
	d	-	-	-
141-6	v	-	-	-
	e	R	R,O	S
	d	R	R,O	S

continued

Table 30. (continued)

TOOL NUMBER		FREQ	ORIENT	SIZE
WOOD (continued)				
141-11	v	R	O,R	S
	e	-	-	-
	d	-	-	-
161-20	v	-	-	-
	e	-	-	-
	d	-	-	-
26-10	v	-	-	-
	e	-	-	-
	d	-	-	-
27-15	v	R	R	S
	e	C	R	S
	d	R	R	S
134-23	v	-	-	-
	e	C	R	S
	d	C	R	S
138-6	v	-	-	-
	e	-	-	-
	d	-	-	-
138-16	v	-	-	-
	e	-	-	-
	d	-	-	-
141-4	v	-	-	-
	e	R	R	S
	d	-	-	-
141-17	v	R	R	R
	e	-	-	-
	d	R	R	R
27-10	v	-	-	-
	e	-	-	-
	d	-	-	-

continued

Table 30. (continued)

TOOL NUMBER		FREQ	ORIENT	SIZE
WOOD (continued)				
134-18	v	C	P,O	S
	e	-	-	-
	d	-	-	-
63-2	v	-	-	-
	e	-	-	-
	d	-	-	-
82-3	v	-	-	-
	e	-	-	-
	d	-	-	-
27-7	v	-	-	-
	e	-	-	-
	d	-	-	-
27-14	v	-	-	-
	e	-	-	-
	d	-	-	-
12-5	v	-	-	-
	e	R	R	S
	d	-	-	-
17-12	v	-	-	-
	e	-	-	-
	d	-	-	-
80-9	v	-	-	-
	e	R	R	S
	d	-	-	-
80-10	v	-	-	-
	e	R	R,O	S
	d	-	-	-
80-11	v	-	-	-
	e	-	-	-
	d	-	-	-

continued

Table 30. (continued)

TOOL NUMBER		FREQ	ORIENT	SIZE
WOOD (continued)				
23-4	v	-	-	-
	e	-	-	-
	d	-	-	-
27-8	v	-	-	-
	e	C	R	S
	d	-	-	-
44-7	v	-	-	-
	e	-	-	-
	d	-	-	-
44-9	v	-	-	-
	e	-	-	-
	d	-	-	-
161-15	v	-	-	-
	e	-	-	-
	d	-	-	-

KEY

v = Ventral aspect of used edge.

e = Used edge.

d = Dorsal aspect of used edge.

FREQ = Striation frequency: "R" rare, "C" common.

ORIENT = Orientation of striations in relation to edge axis of tool: "R" right angles to edge axis, "O" oblique to edge axis, "P" parallel to edge axis.

SIZE = Length of striations: "S" short, "M" medium, "L" long.

APPENDIX B

LIGHT MICROSCOPE PHOTOMICROGRAPHS OF MICROWEAR

Note: Magnifications in the figure captions include a photo-enlargement factor of 3.5 x. Photomicrographs were taken at the magnifications listed in the left hand column below. Printing the photomicrographs with a 3.5 x enlargement of the image on the negative resulted in the magnifications listed in the right hand column below.

100 x	350 x
200 x	700 x
250 x	875 x
300 x	1050 x
400 x	1400 x
600 x	2100 x

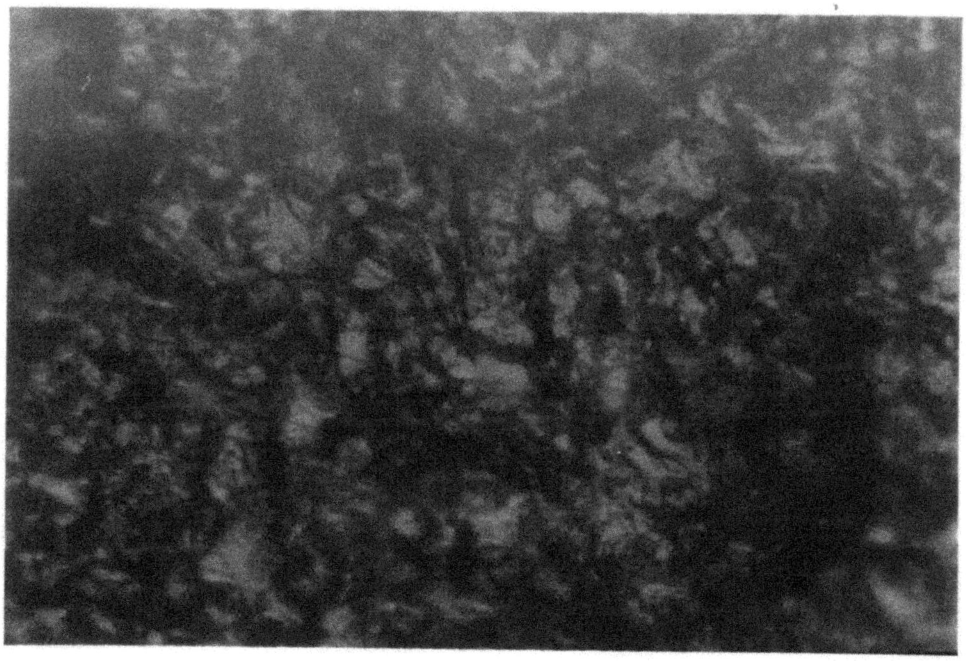

Figure 32. Unused ventral surface of tool 134-23. (700 x)

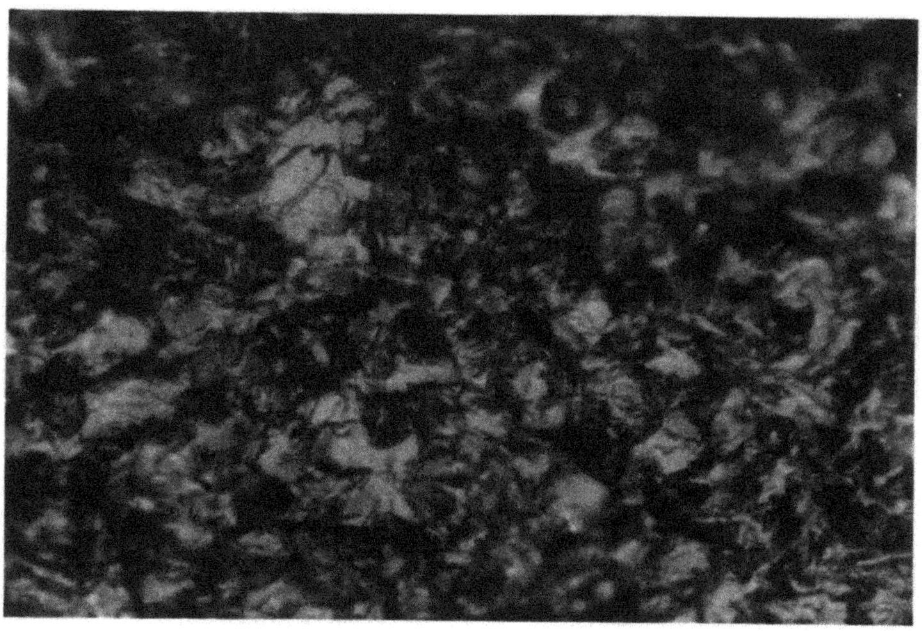

Figure 33. Unused dorsal surface of tool 27-17. (1050 x)

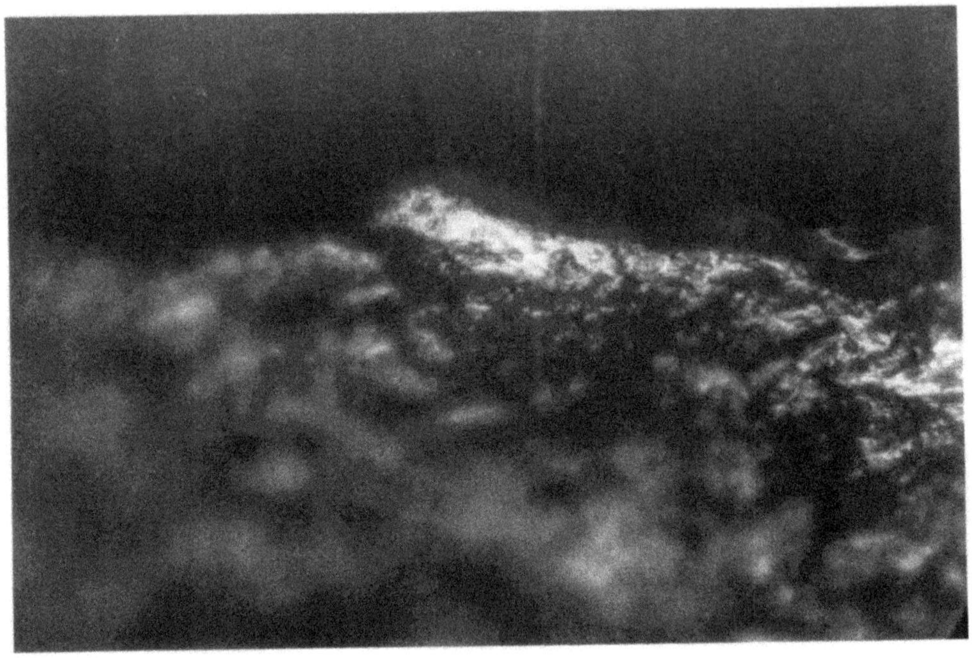

Figure 34. Bright, smooth-pitted polish on a projecting part of the edge and dorsal edge-aspect of tool 134-11. The tool was used to cut plants (1000 strokes).
(700 x)

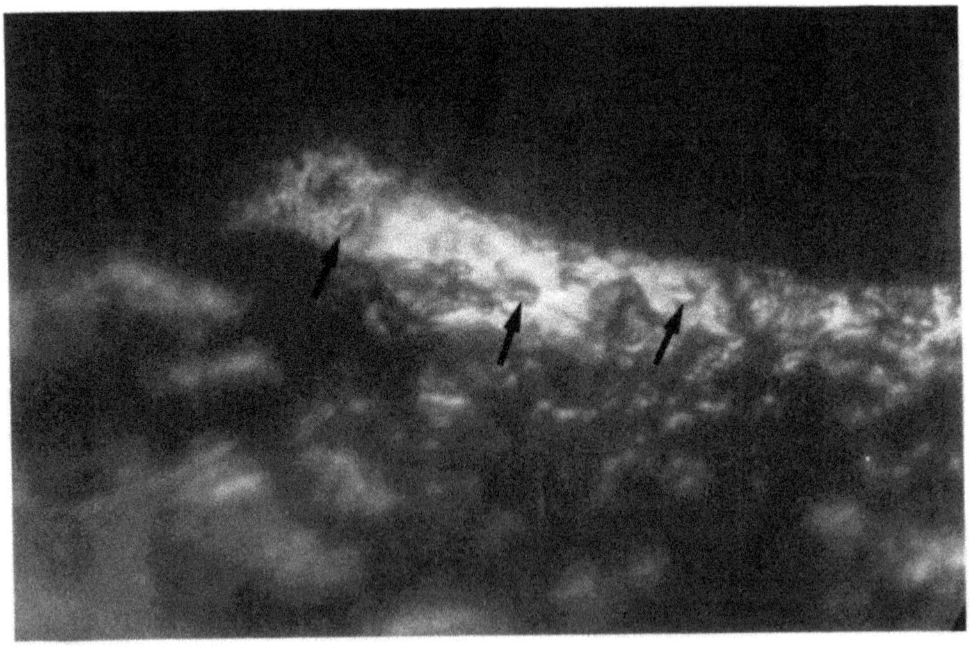

Figure 35. Close-up view of same area of polish on tool 134-11 as shown in *Figure 34.* Arrows point to pits in polish.
(1400 x)

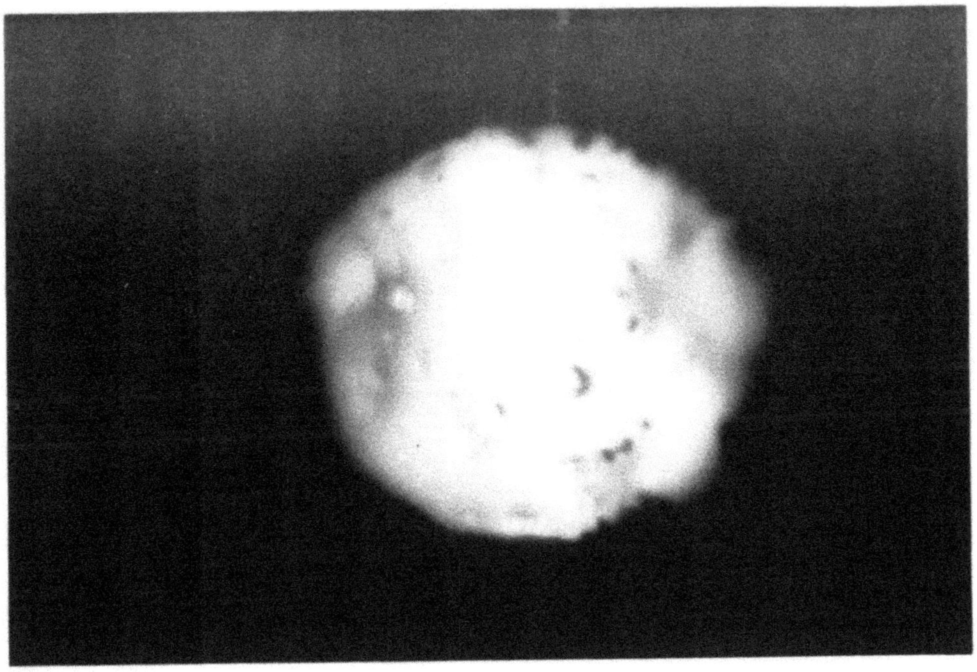

Figure 36. Bright, smooth-pitted polish on the ventral edge-aspect of tool 161-11. The tool was used to cut plants (700 strokes).
(1400 x)

Figure 37. Dull, grainy polish on the dorsal edge-aspect of tool 17-14. The tool was used to dehair hide (5200 strokes).
(1050 x)

Figure 38. Close-up of a striation (arrowhead) on tool 17-14, also shown in *Figure 37*.
(2100 x)

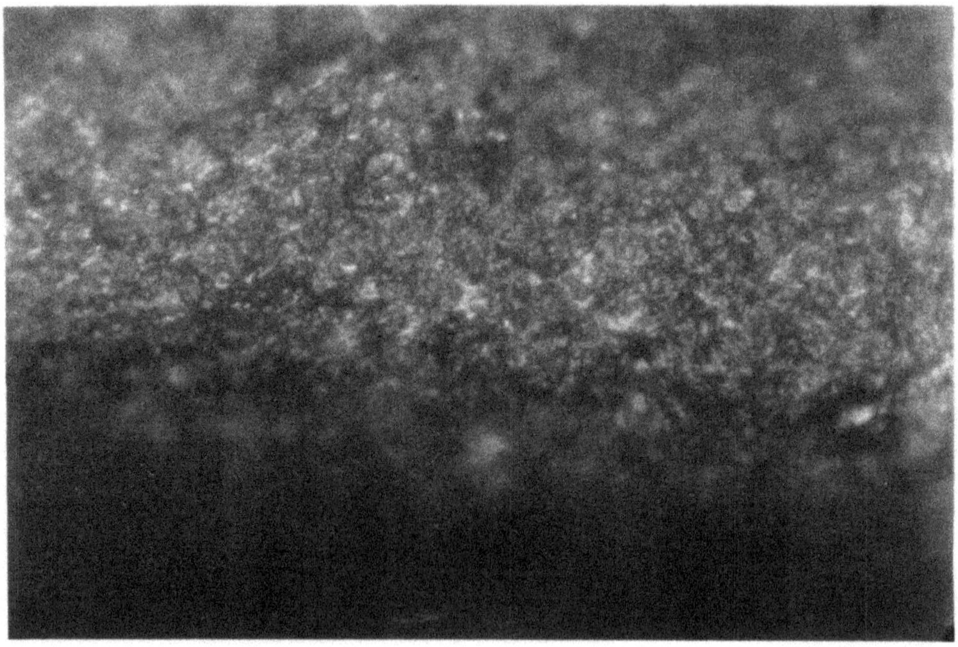

Figure 39. Dull, grainy polish on the dorsal edge-aspect and edge of tool 12-13. The tool was used to dehair hide (4000 strokes).
(700 x)

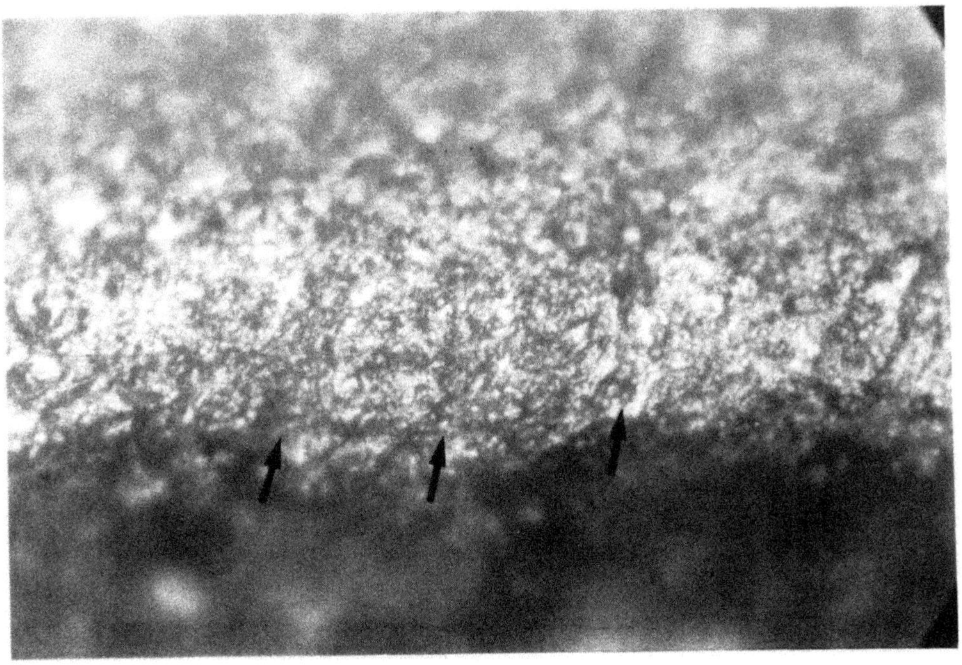

Figure 40. Dull, grainy polish on a different part of the dorsal edge-aspect and edge of tool 12-13 than shown in *Figure 39*. Arrows point to trend lines in the polish.
(700 x)

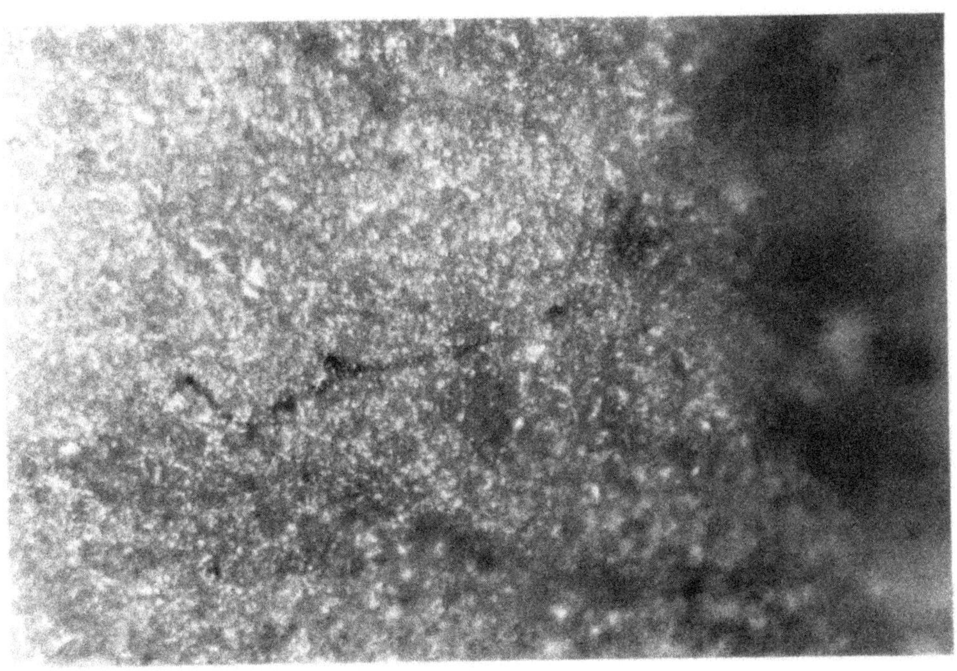

Figure 41. Dull, grainy polish on the dorsal edge-aspect of tool 27-17. The tool was used to deflesh hide (2500 strokes).
(350 x)

Figure 42. Dull, grainy polish of a different part of the dorsal edge-aspect of tool 27-17 than shown in *Figure 41.*
(700 x)

Figure 43. Close-up view of the dull, grainy polish on tool 27-17 shown in *Figure 42.*
(1400 x)

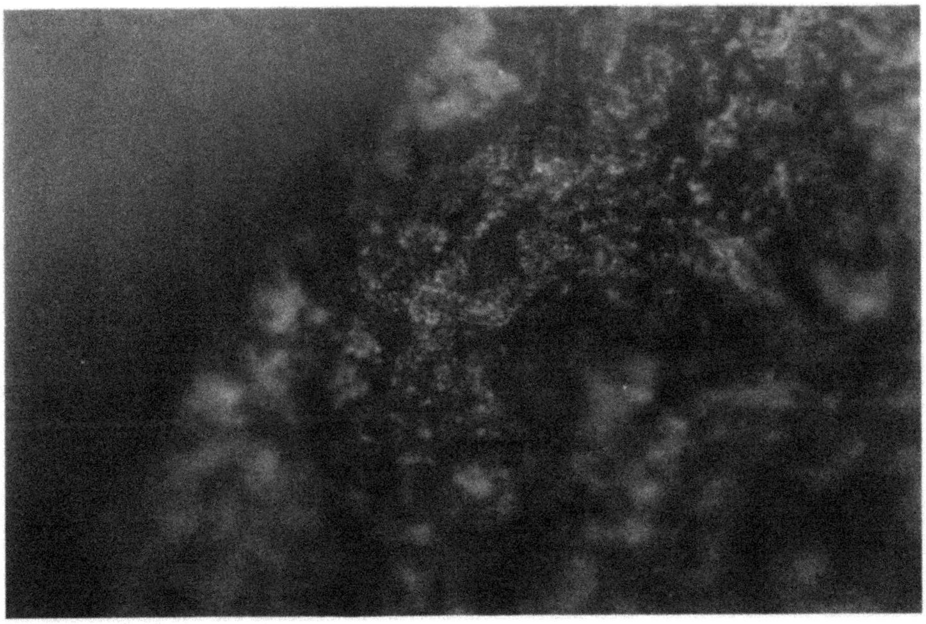

Figure 44. Dull, grainy polish on a high ridge of the ventral edge-aspect near the edge of tool 134-9. The tool was used to saw through hide (1200 strokes).
(700 x)

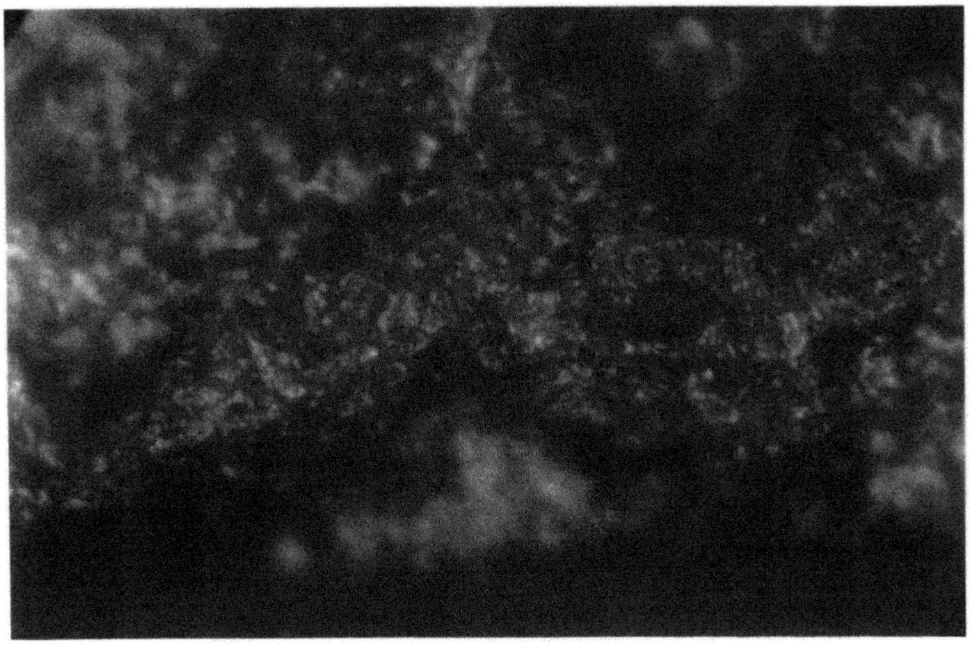

Figure 45. Slightly different view of the dull, grainy polish on tool 134-9 shown in *Figure 44*.
(875 x)

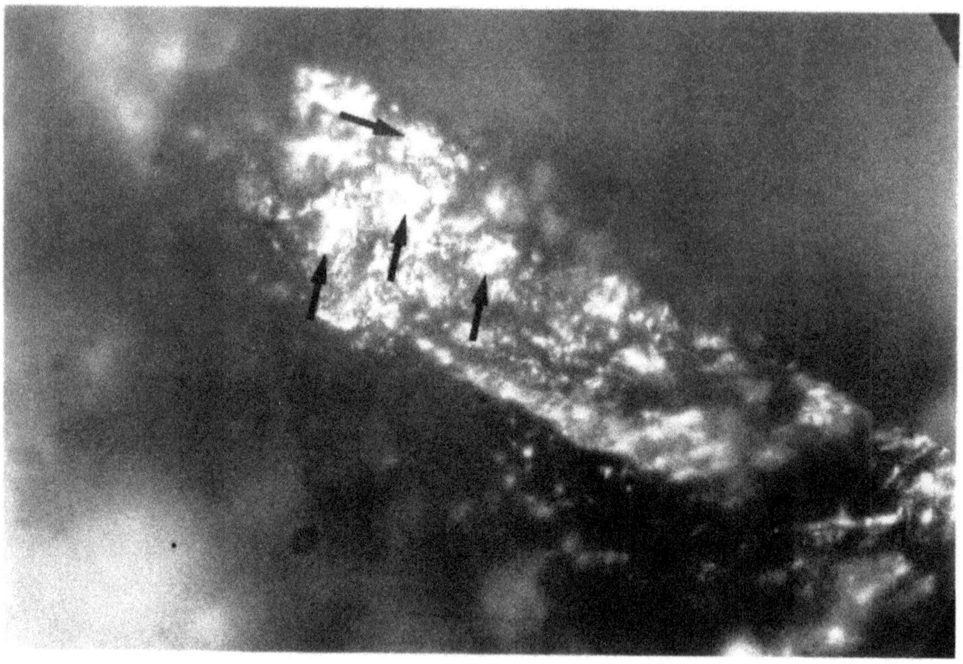

Figure 46. Dull to bright, grainy polish with bright, smooth-pitted areas (arrows) on the ventral edge-aspect of tool 46-2. The tool was used to saw bone (400 strokes).
(700 x)

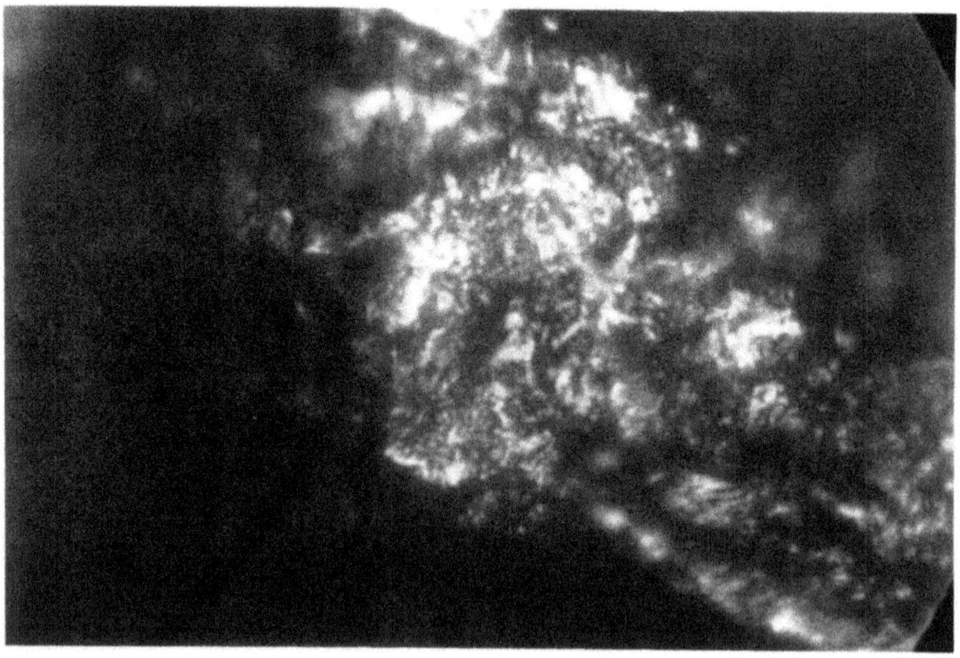

Figure 47. Close-up view of the polished area of tool 46-2 shown in *Figure 46.*
(1400 x)

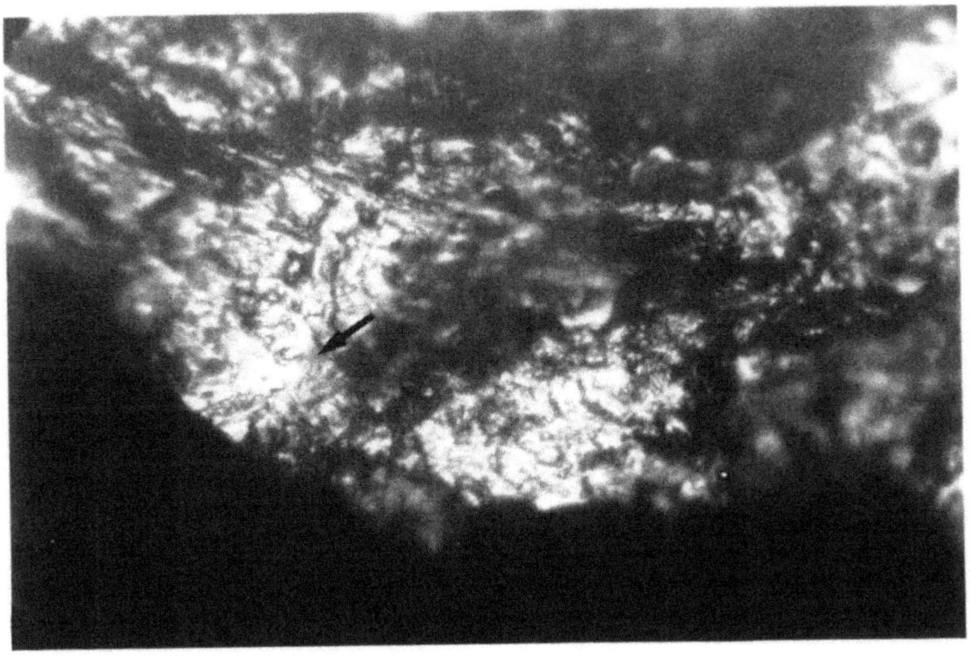

Figure 48. New area on the ventral edge-aspect of tool 46-2. Smooth-pitted polish to left (arrow) and bright, smooth to grainy polish to center-right.
(700 x)

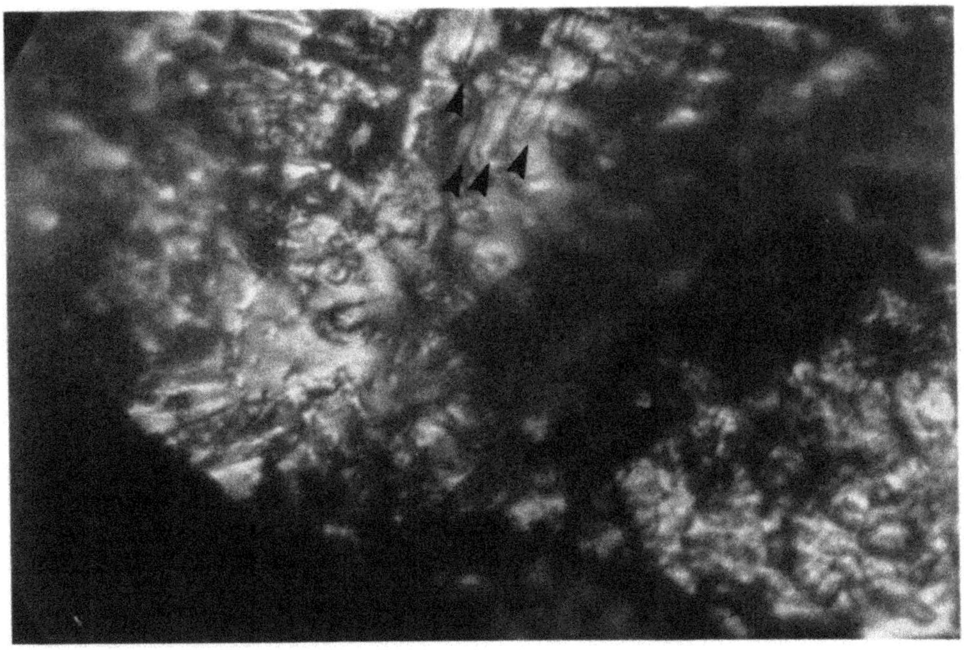

Figure 49. Close-up view of the smooth-pitted polish on tool 46-2 shown in *Figure 48*. Possible striations (arrowheads) are oriented at oblique angles to the edge of the tool.
(1400 x)

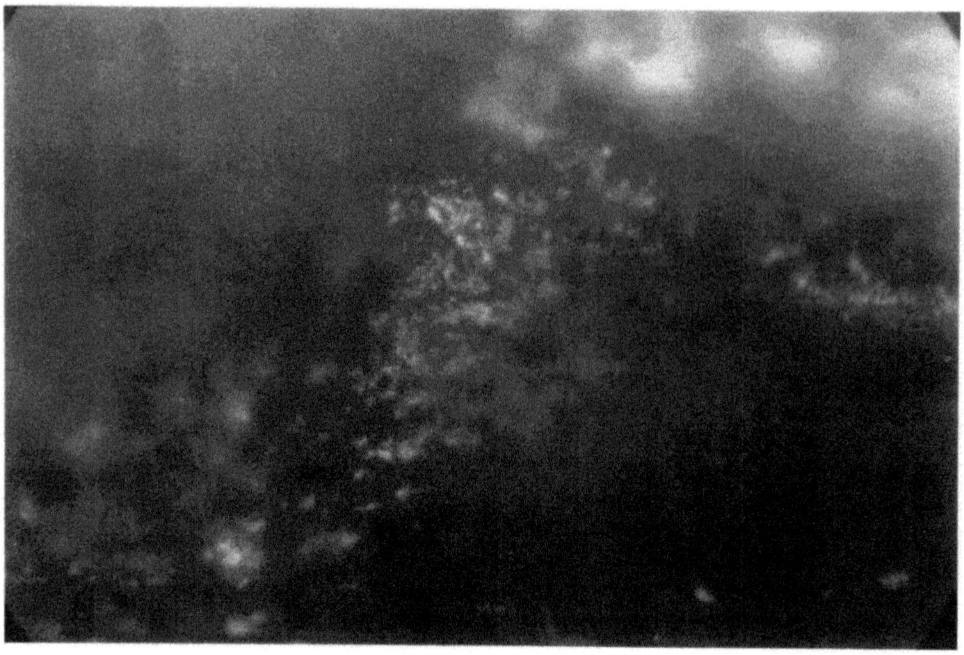

Figure 50. Dull, grainy polish with a few bright, smooth spots on the dorsal edge-aspect of tool 17-13. The tool was used to scrape bone (100 strokes).
(700 x)

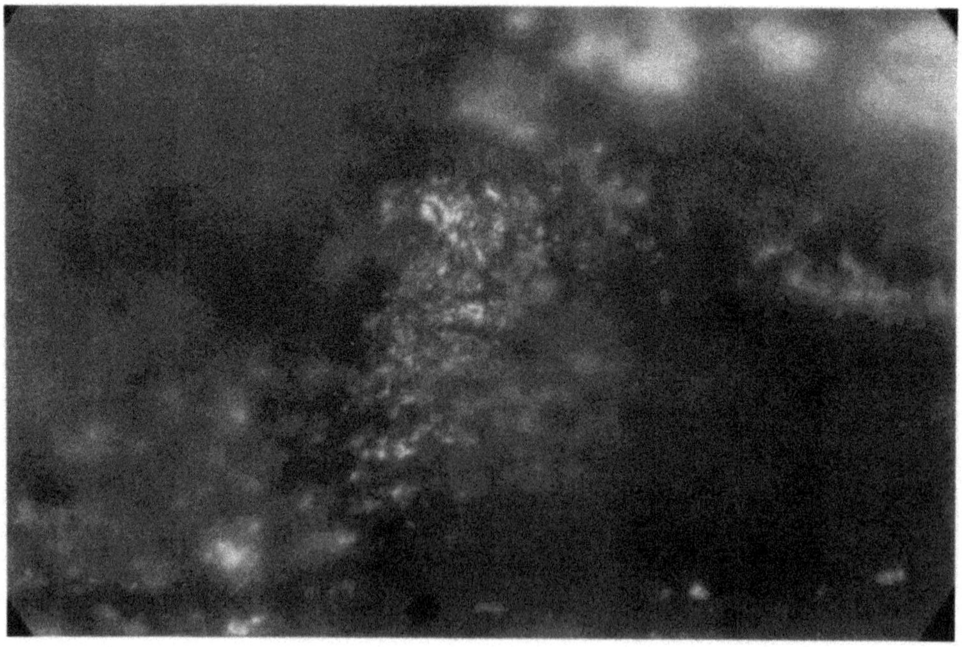

Figure 51. Close-up view of the polish on tool 17-13 shown in *Figure 50.*
(875 x)

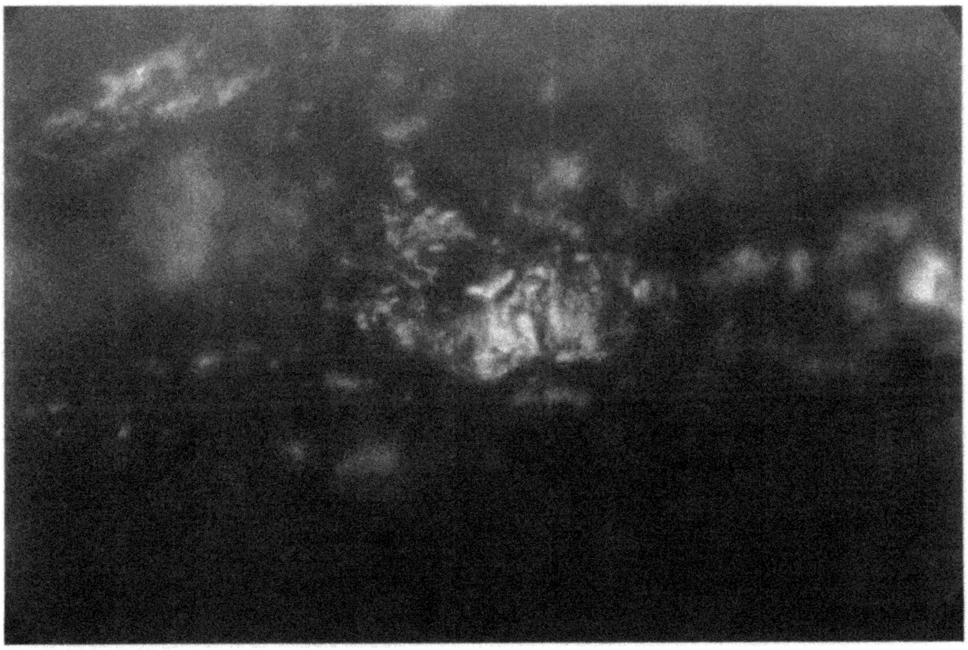

Figure 52. Bright, smooth to grainy polish on a different part of the dorsal edge-aspect of tool 17-13 to that shown in *Figures 50* and *51*.
(875 x)

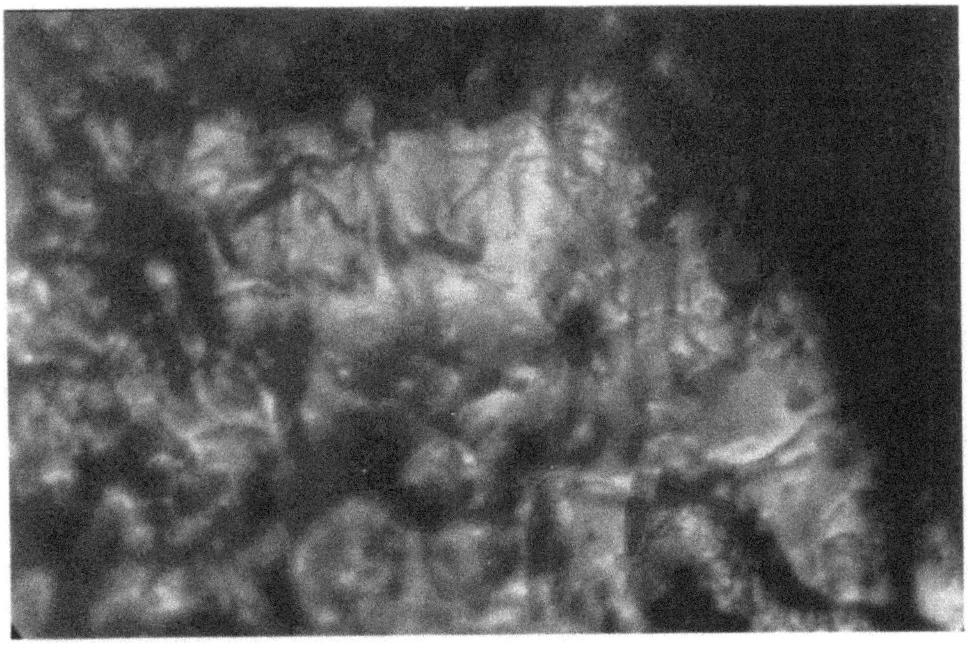

Figure 53. Bright, smooth-pitted polish on the dorsal edge-aspect near the tip of tool 138-19. The tool was used to engrave bone (600 strokes).

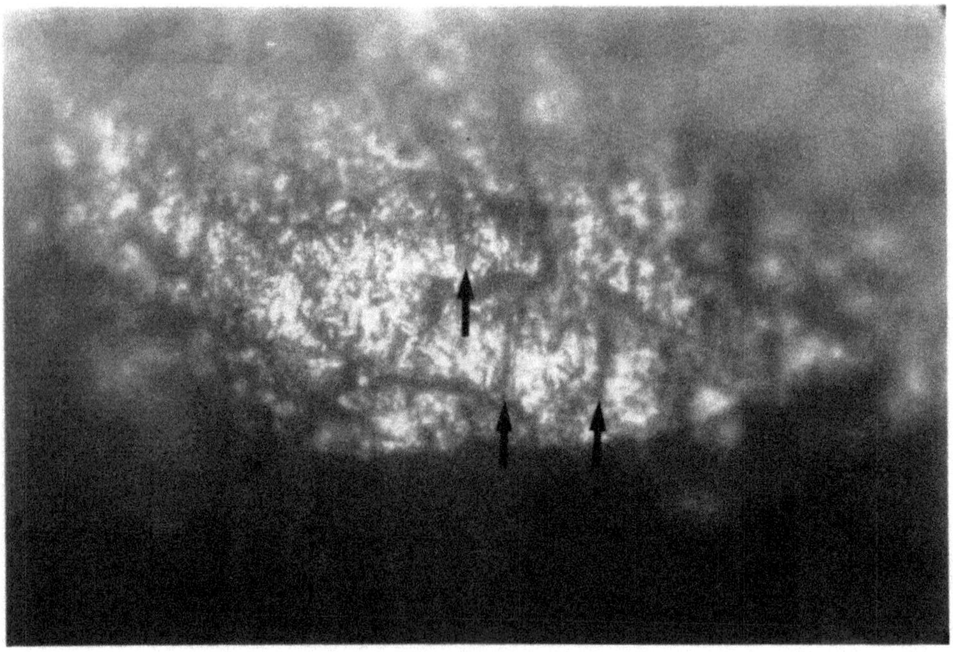

Figure 54. Smooth-pitted polish surrounded by bright, grainy and dull, grainy polish with trend lines (arrows) near the tip of tool 134-6. The tool was used to bore into antler (1000 strokes). (700 x)

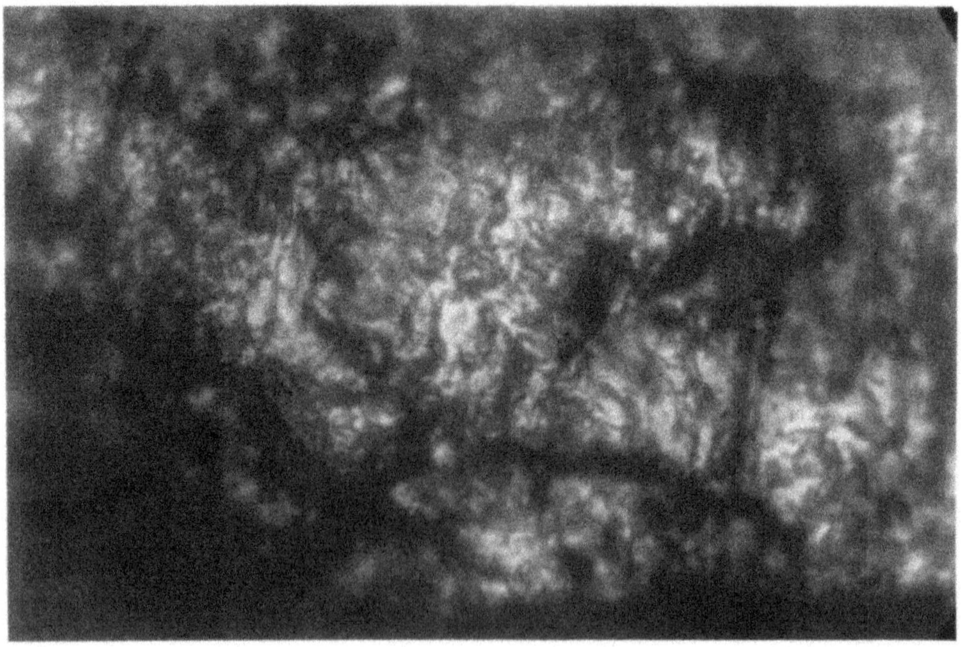

Figure 55. Close-up view of polish on tip of tool 134-6 shown in the center of *Figure 54.*
(1400 x)

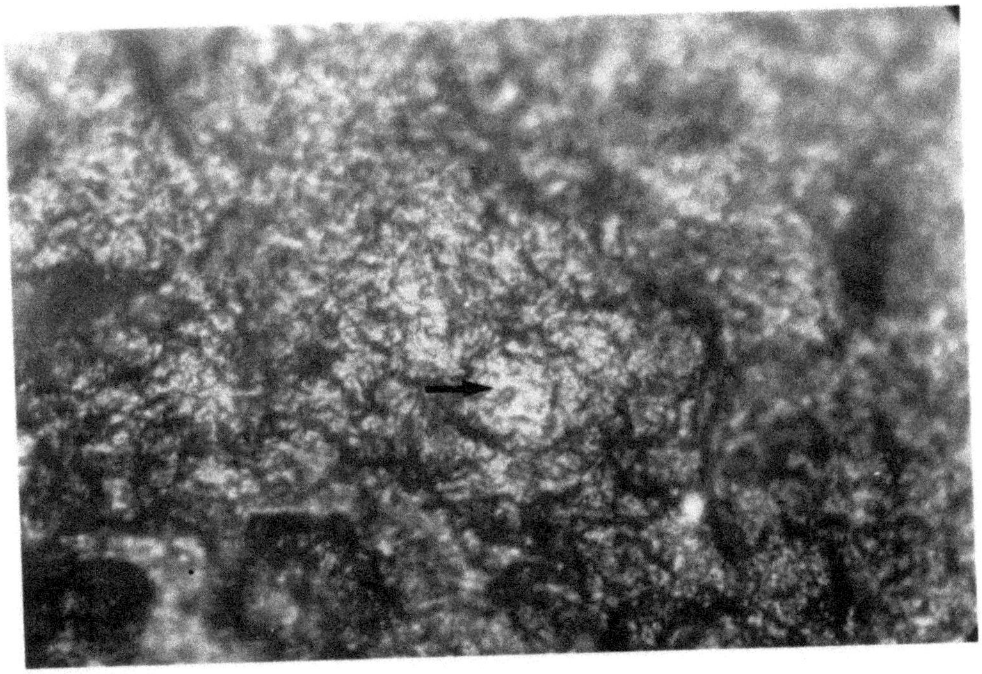

Figure 56. Dorsal edge-aspect of tool 134-6. Bright, grainy polished area with a smooth-pitted core (arrow) on a high point. Dull, to bright, grainy polish near the bottom of photo. (700 x)

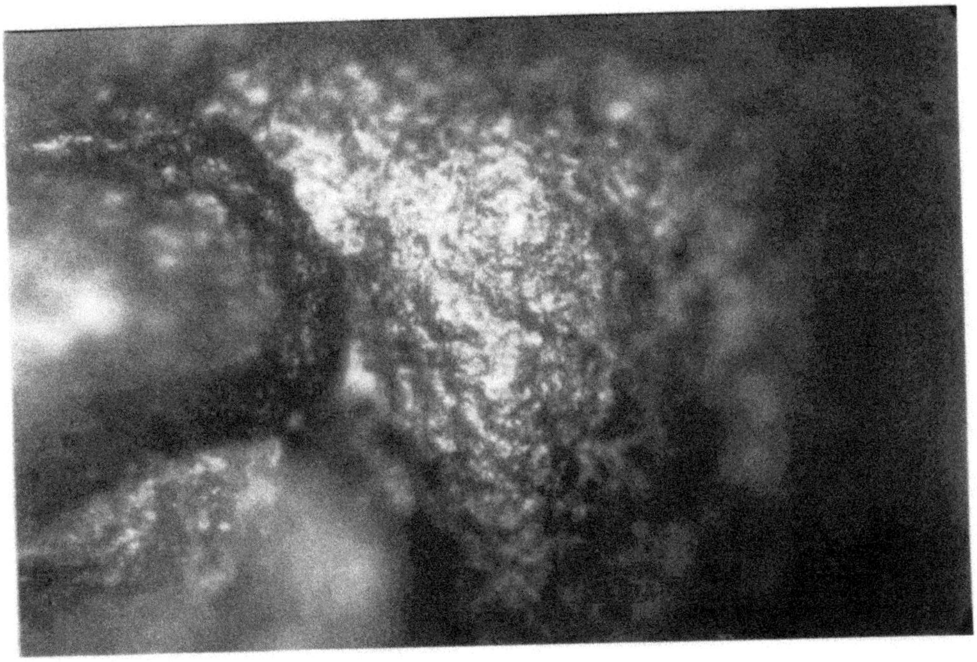

Figure 57. Bright, grainy polish with a few bright, smooth spots on a spall from the tip of tool 141-20, detached after approximately 800 engraving strokes on antler.

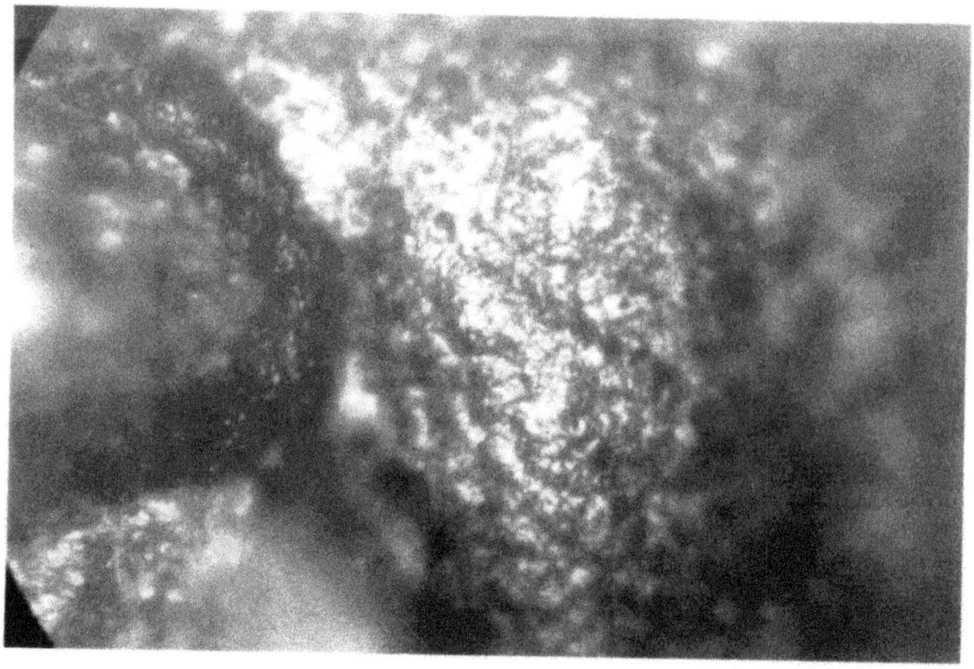

Figure 58. Close-up view of polish on tool 141-20 shown in *Figure 57.*
(875 x)

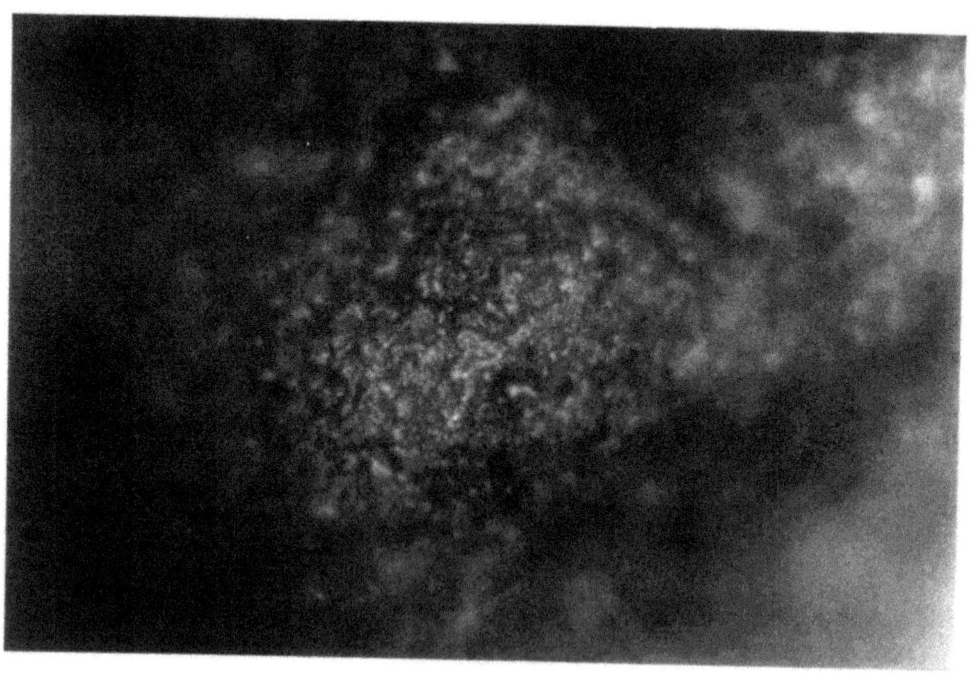

Figure 59. Dull, grainy polish on the dorsal edge-aspect of a new area away from the tip of tool 141-20. The tool was used for 1000 strokes to engrave antler.

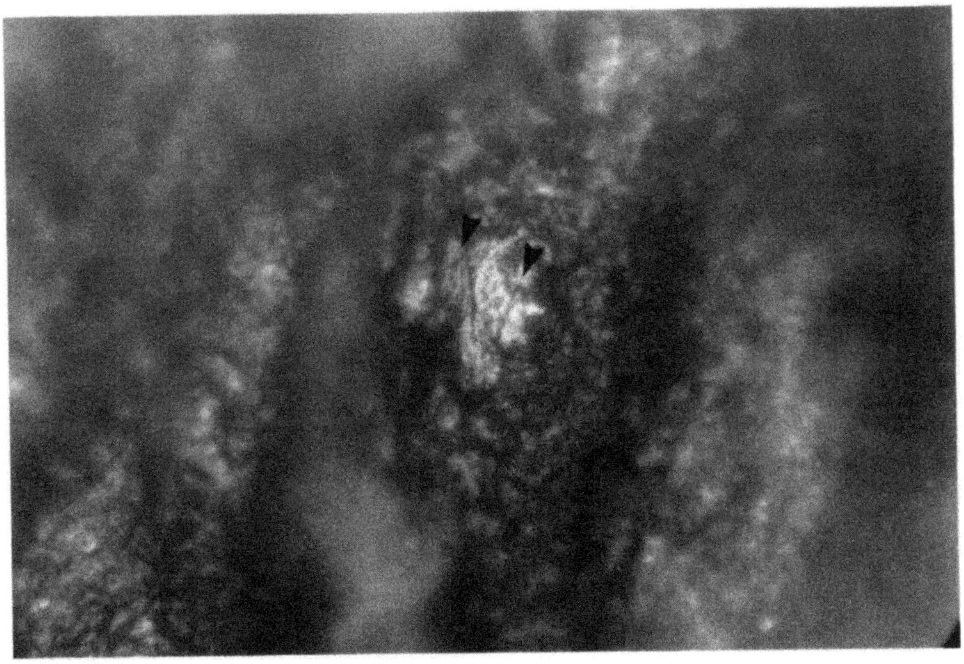

Figure 60. New area near the tip of tool 141-20. Bright, smooth-pitted polish on a high point surrounded by dull, grainy polish. Note striations (arrowheads) in the smooth-pitted polish. (700 x)

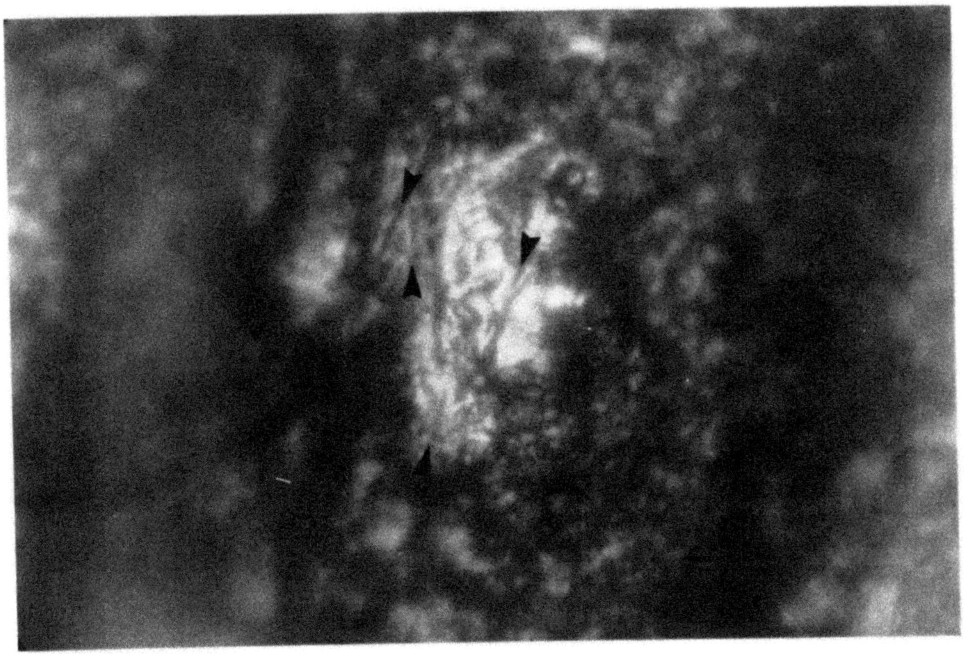

Figure 61. Close-up view of the bright, smooth-pitted polish on tool 141-20 shown in *Figure 60.* Arrowheads point to striations. (1400 x)

Figure 62. Dull to bright, grainy polish (central area of photo) on the dorsal edge-aspect of tool 12-3. The tool was used to eviscerate a deer (750 strokes).
(700 x)

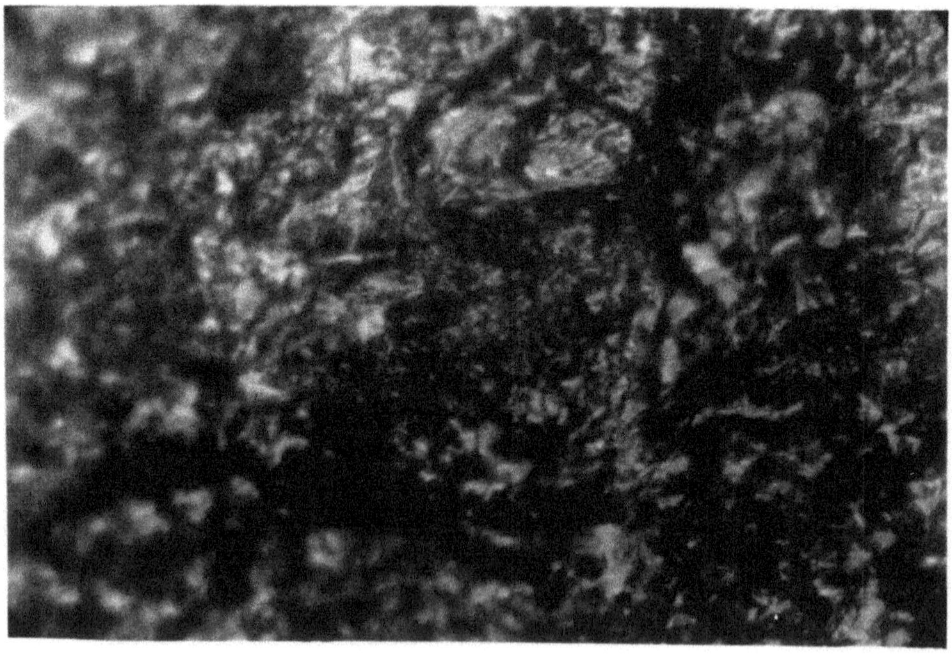

Figure 63. Dull to bright, grainy polish (centre of photo) on the ventral edge-aspect of tool 45-7. The tool was used to cut flesh (2000 strokes).

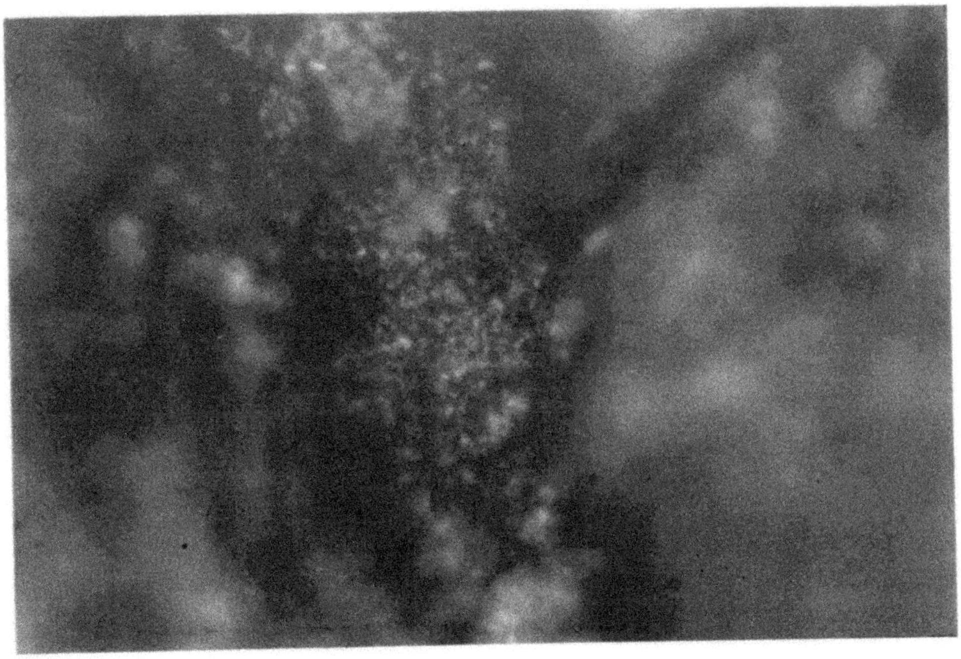

Figure 64. Dull to bright, grainy polish on a ridge on the ventral edge-aspect of tool 27-5. The tool was used to cut flesh (150 strokes).
(700 x)

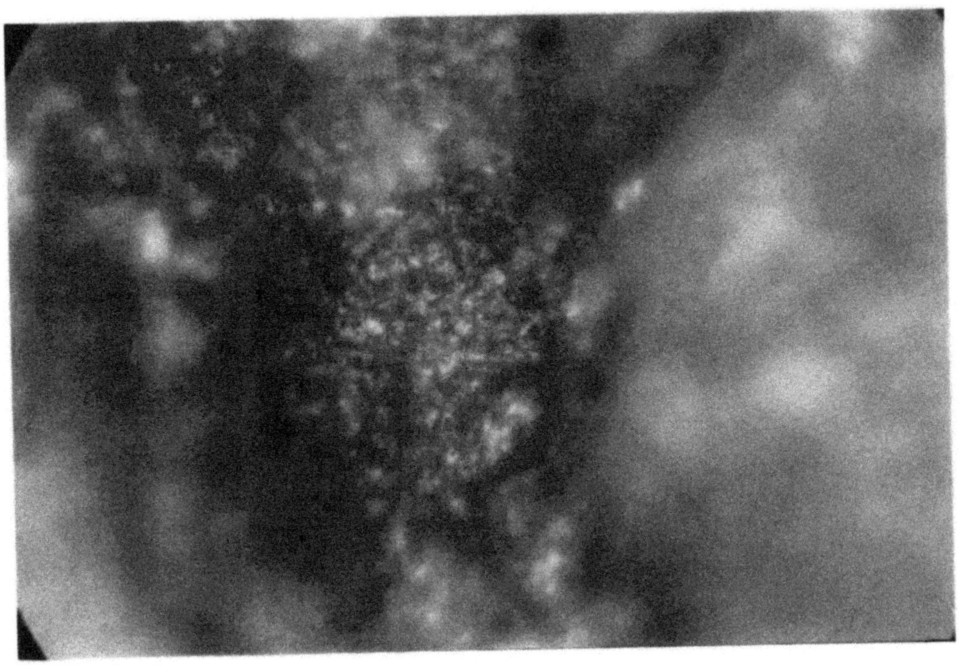

Figure 65. Close-up view of polish on tool 27-5 shown in *Figure 64.*
(875 x)

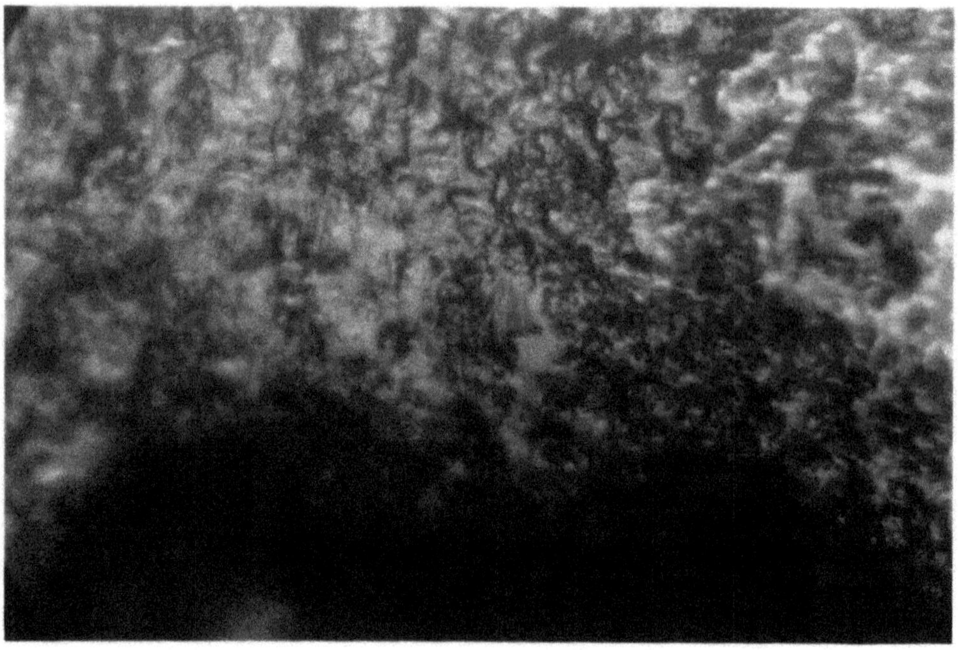

Figure 66. Smooth-pitted polish and dull-bright, grainy polish on the dorsal edge-aspect and edge of tool 134-23. The tool was used to scrape wood (5850 strokes).
(1050 x)

Figure 67. Close-up view of polish on tool 134-23 shown in *Figure 66.* Arrowheads point to striations.
(1400 x)

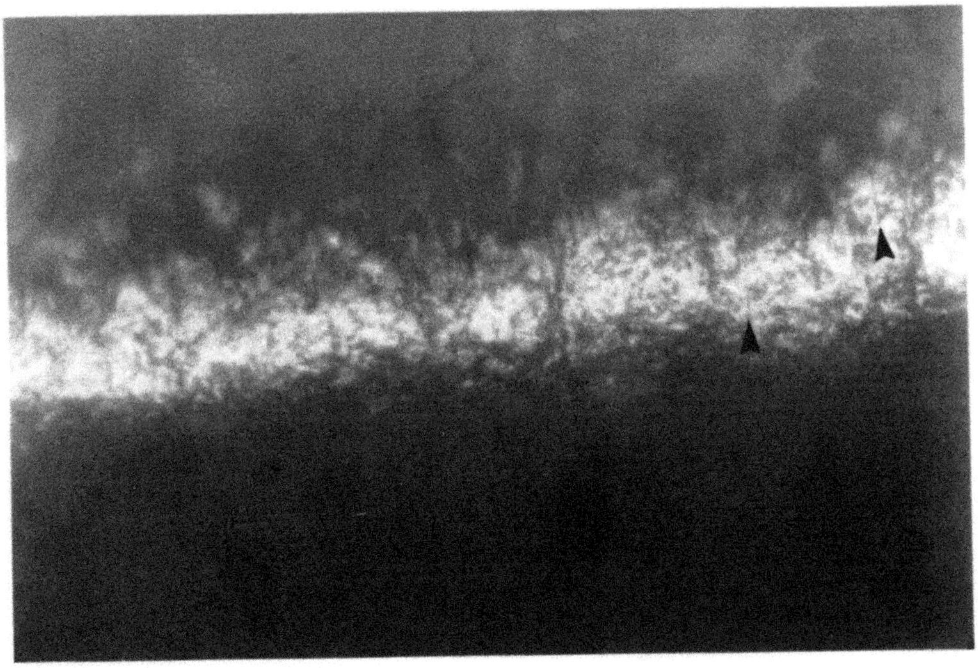

Figure 68. Smooth-pitted polish and striations (arrowheads) on the rounded edge of a new area on tool 134-23.
(700 x)

Figure 69. Close-up view of polish and striations on tool 134-23 shown in center of *Figure 68.*
(1400 x)

APPENDIX C

SCANNING ELECTRON MICROSCOPE PHOTOMICROGRAPHS OF MICROWEAR

Note: Scale bar along bottom of photomicrographs and scale interval on the left end of the data line.

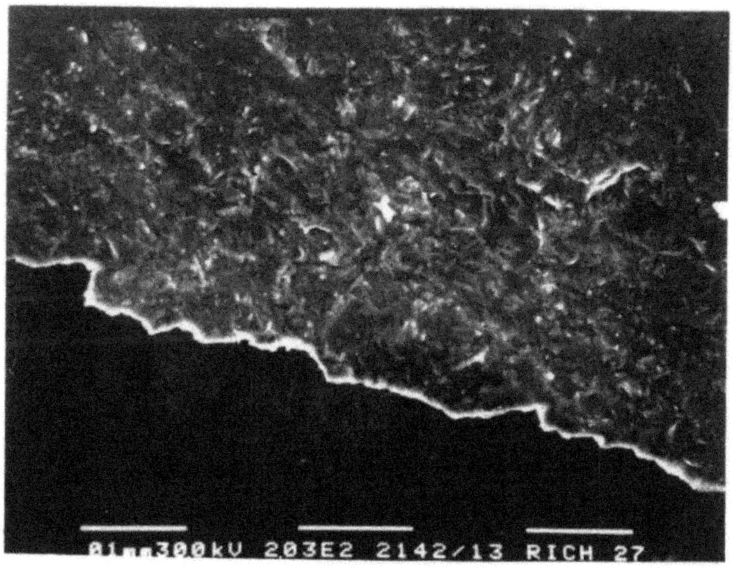

Figure 70. Unused vitreous basalt tool. (203 x)

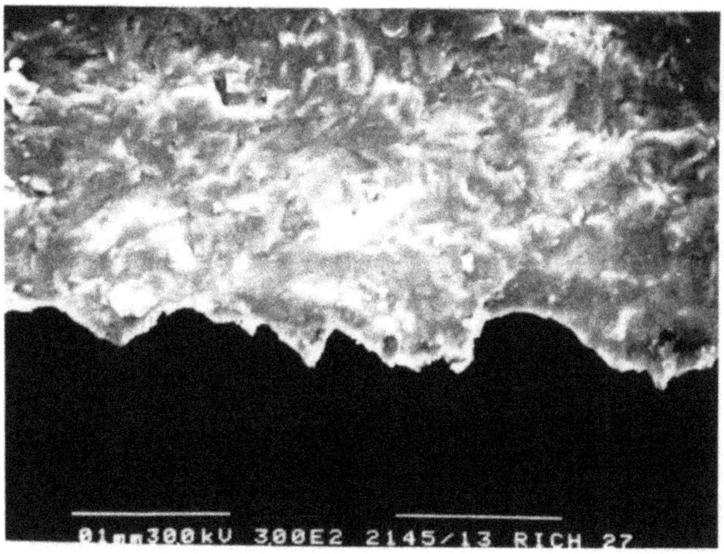

Figure 71. Unused vitreous basalt tool. (300 x)

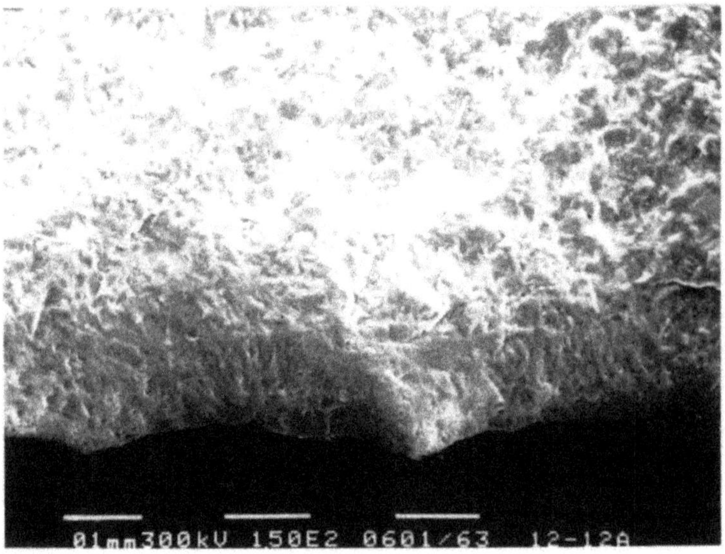

Figure 72. Continuous, smooth-pitted polish along the edge and dorsal edge aspect of tool 12-12. The tool was used to cut plants (1750 strokes).
(150 x)

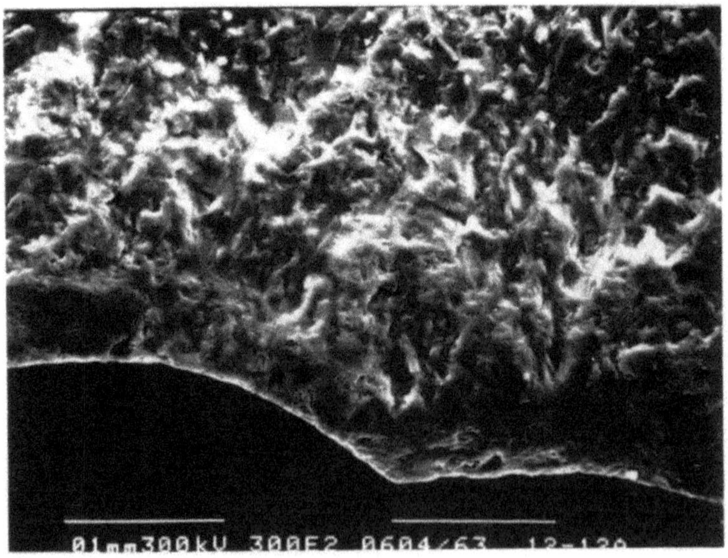

Figure 73. Smooth-pitted polish on a new area of the edge and dorsal edge-aspect of tool 12-12.
(300 x)

Figure 74. Close-up view of polish on the edge and dorsal edge-aspect of tool 12-12. Note smooth texture of the polish, the deep pits (arrows) and short, shallow striations (arrowheads) oriented parallel or oblique to the edge.
(1250 x)

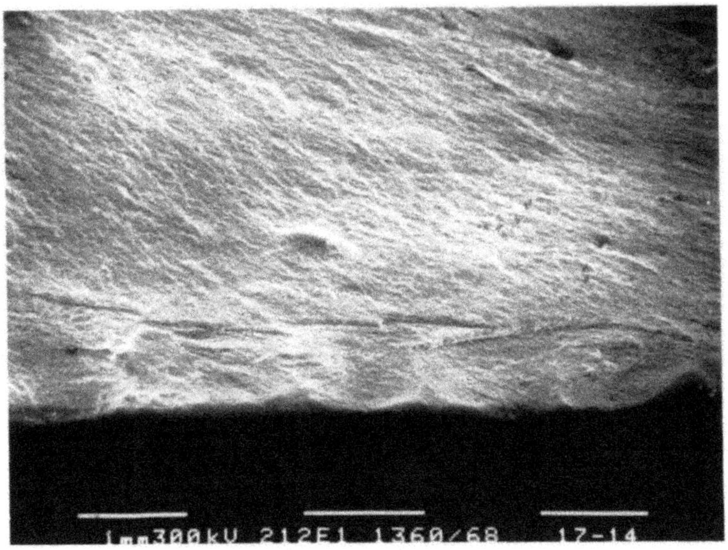

Figure 75. Step-terminated microflake scars on the dorsal edge-aspect and rounded edge of tool 17-14. The tool was used to scrape deer hide (5200 strokes).
(21.2 x)

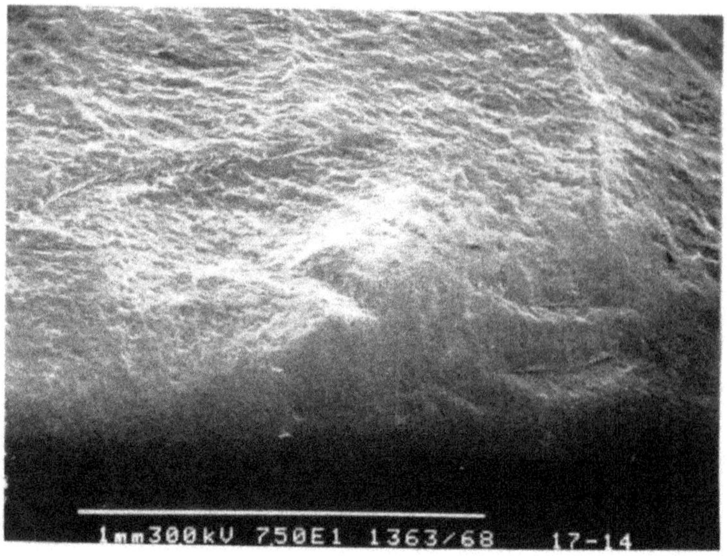

Figure 76. New area of the rounded edge and dorsal edge-aspect of tool 17-14.
(75 x)

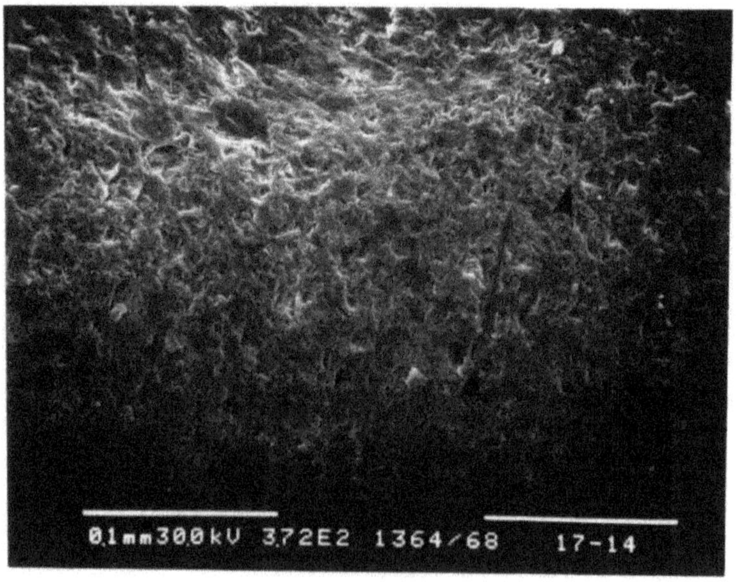

Figure 77. Polish on the rounded edge of tool 17-14 visible in the center of *Figure 76.* Striations (arrowheads) oriented at approximate right-angles to the edge are visible on the "platy" textured polish.
(372 x)

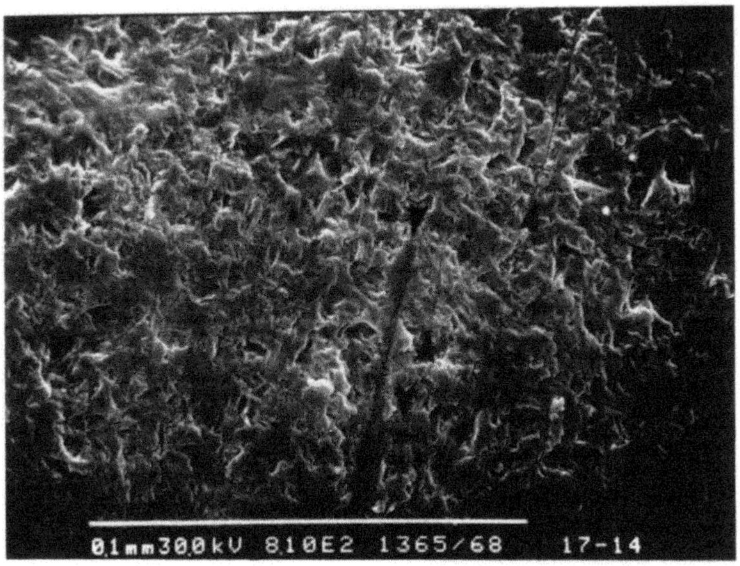

Figure 78. Close-up view of striations (arrowheads) on tool 17-14 shown in *Figure 77.* Note additional short, narrow striations which were not visible at lower magnifications.
(810 x)

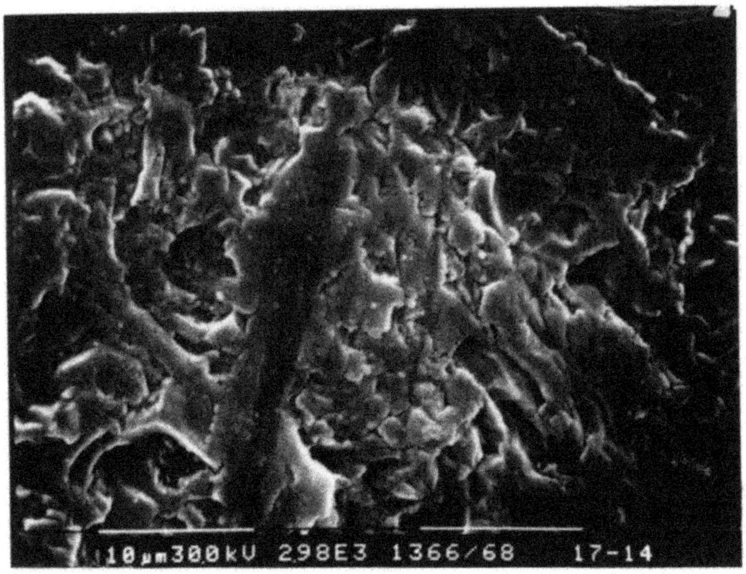

Figure 79. Extreme close-up view of the upper part of the large striation shown in *Figure 78.* At this magnification it appears to be comprised of several parallel scratches.
(2980 x)

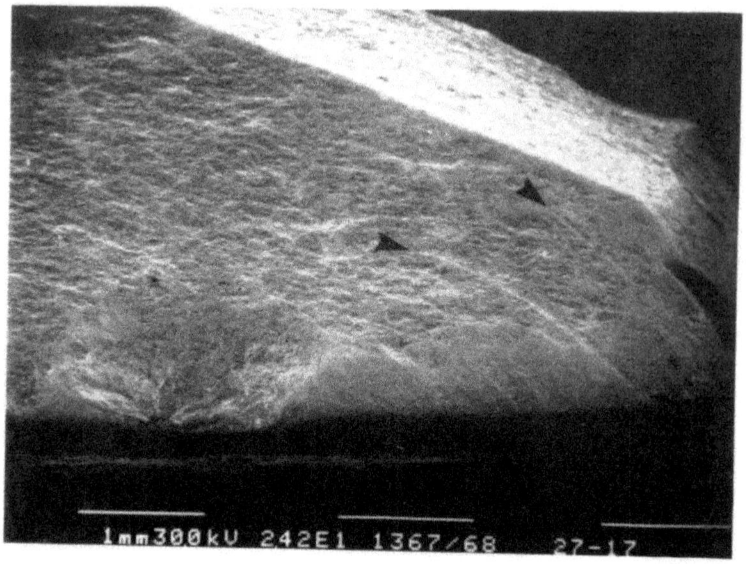

Figure 80. Extensively rounded and polished edge and dorsal edge-aspect of tool 27-17. Note two large striations (arrowheads) arcing from the polished surface onto the unpolished surface. The tool was used to scrape deer hide (2500 strokes).
(24.2 x)

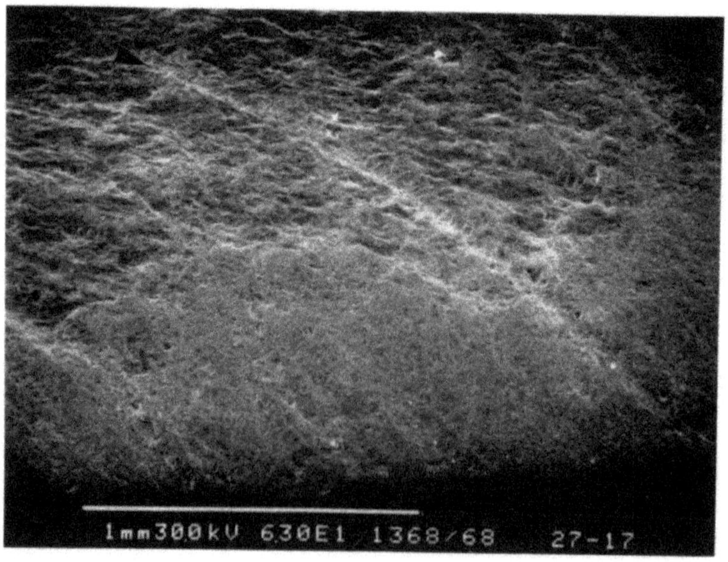

Figure 81. Close-up view of the polished-unpolished surface boundary and large striation (arrowhead) on tool 27-17 shown in *Figure 80*.
(63 x)

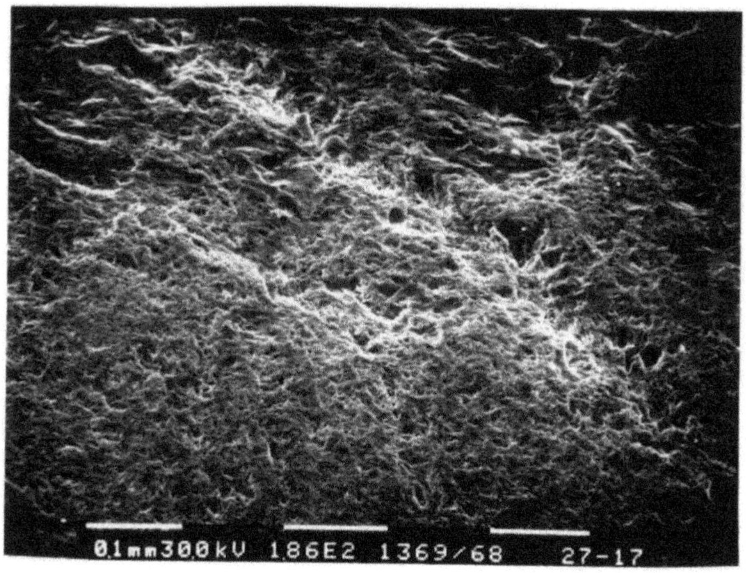

Figure 82. Close-up view of the large striation and the polished-unpolished surface boundary on tool 27-17 shown in *Figures 80* and *81*.
(186 x)

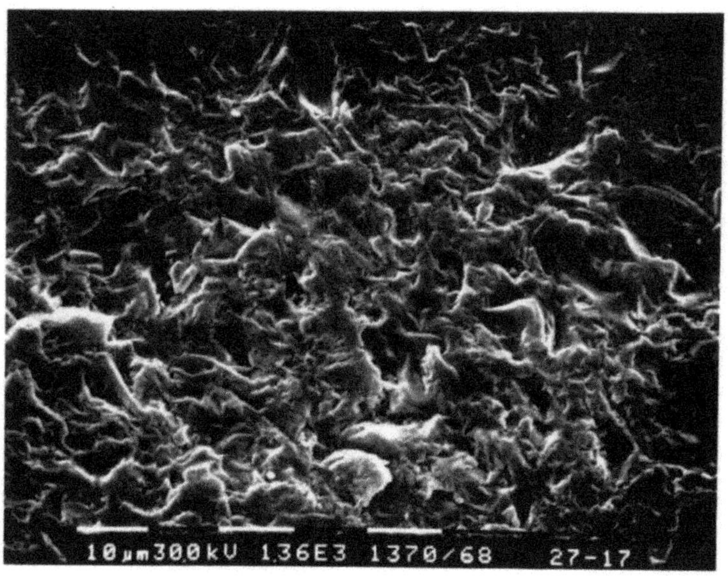

Figure 83. View of "platy" polish and shallow, narrow, parallel striations (arrowhead) on the dorsal edge-aspect of tool 27-17.
(1360 x)

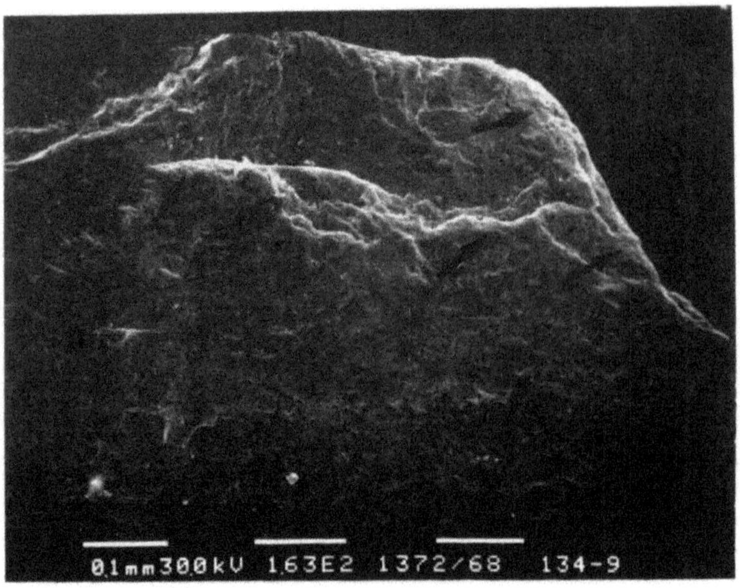

Figure 84. Arrows point to rounding and polish on the edge and backs of microflake scars on tool 134-9. The tool was used to saw through deer hide (1200 strokes).
(163 x)

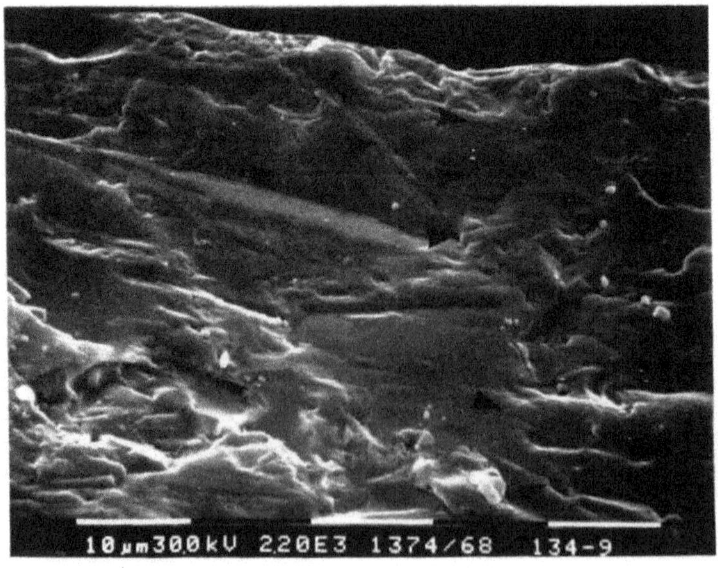

Figure 85. Extensive area of smooth polish on the ventral edge-aspect of tool 134-9. Numerous narrow, shallow striations (arrowheads) are visible in the polish.
(2200 x)

Figure 86. Smooth polish with a tendency to a "platy" texture and possible striations (arrowhead) on the dorsal edge-aspect of tool 134-9.
(2500 x)

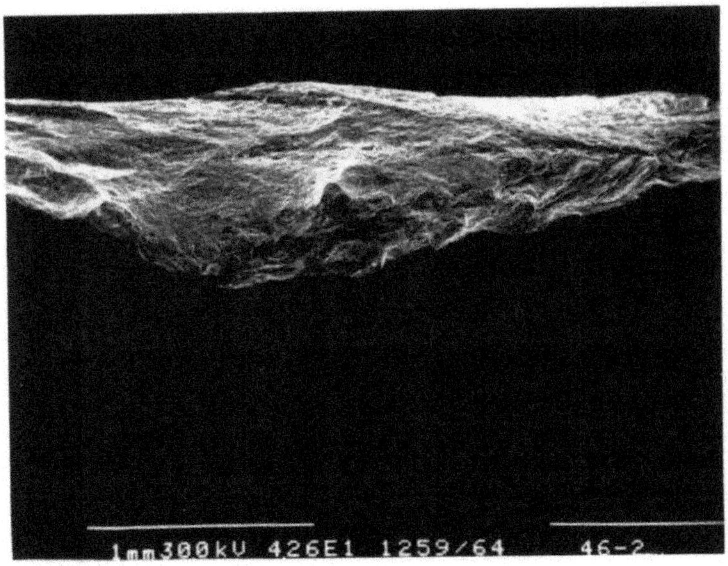

Figure 87. Step- and hinge-terminated microflake scars within a large snap-terminated scar on tool 46-2. The tool was used to saw bone (400 strokes).
(42.6 x)

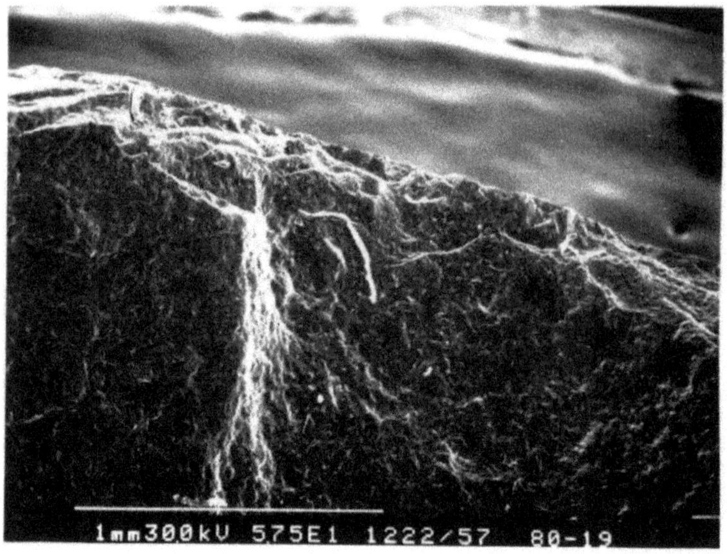

Figure 88. Regularization of the edge and numerous step-terminated microflake scars on the ventral edge-aspect of tool 18-19. The tool was used to saw bone (200 strokes).
(57.5 x)

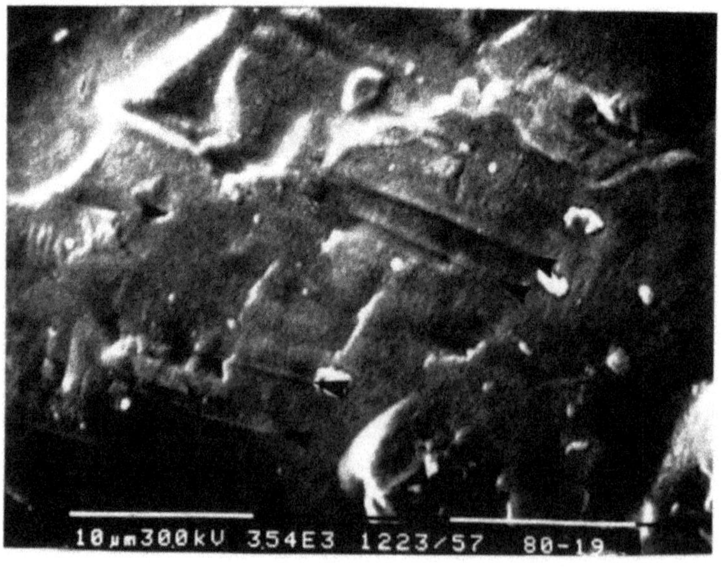

Figure 89. Extreme close-up view of a new area on the ventral edge-aspect of tool 18-19. Note possible striations (arrowheads) visible on a smooth, possibly polished surface.
(3540 x)

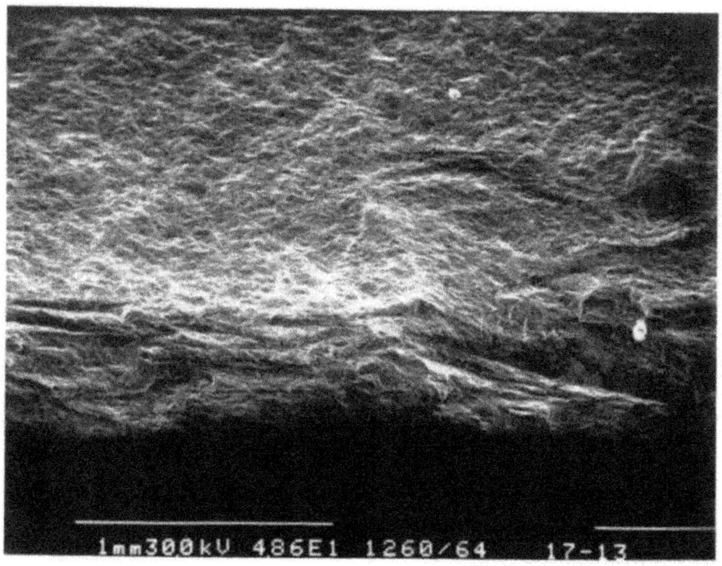

Figure 90. Hinge- and step-terminated microflake scars on the dorsal edge-aspect of tool 17-13. The tool was used to scrape bone (100 strokes).
(48.6 x)

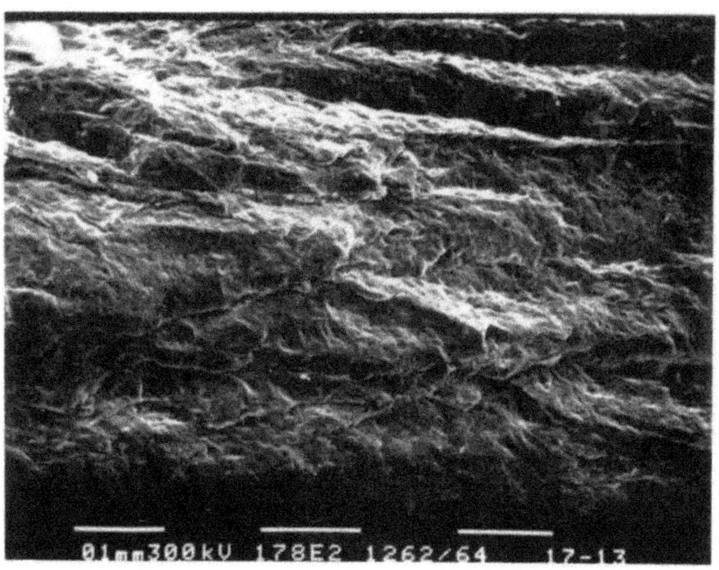

Figure 91. Close-up view of microflake scars on tool 17-13 visible on the far left of *Figure 90.*
(178 x)

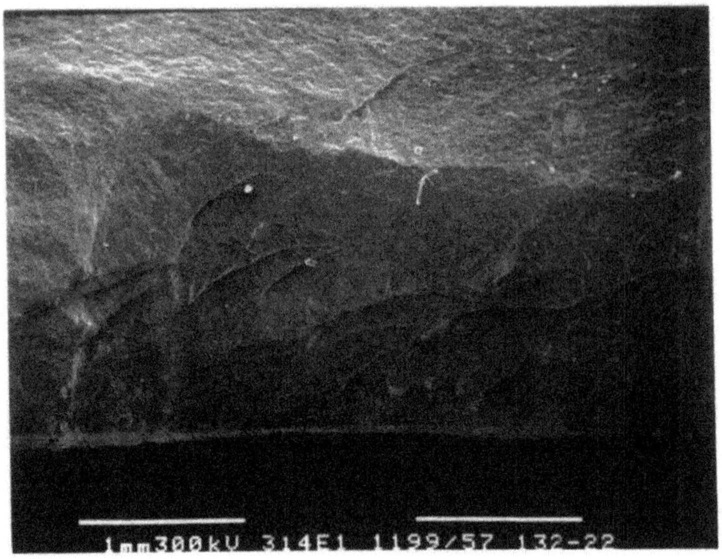

Figure 92. Step-terminated microflake scars within a very large snap-terminated scar on tool 134-22. The tool was used to saw antler (400 strokes).
(31.4 x)

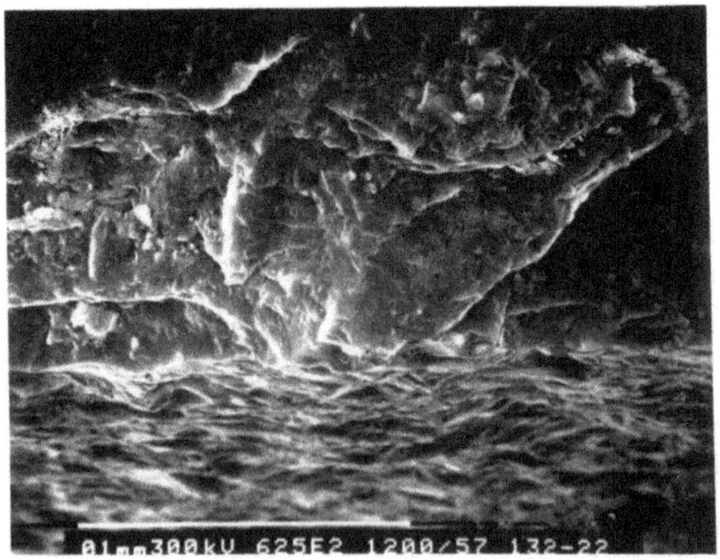

Figure 93. Close-up view of microflake scars on the dorsal edge-aspect of tool 134-22 visible in the center-right of *Figure 92.*
(625 x)

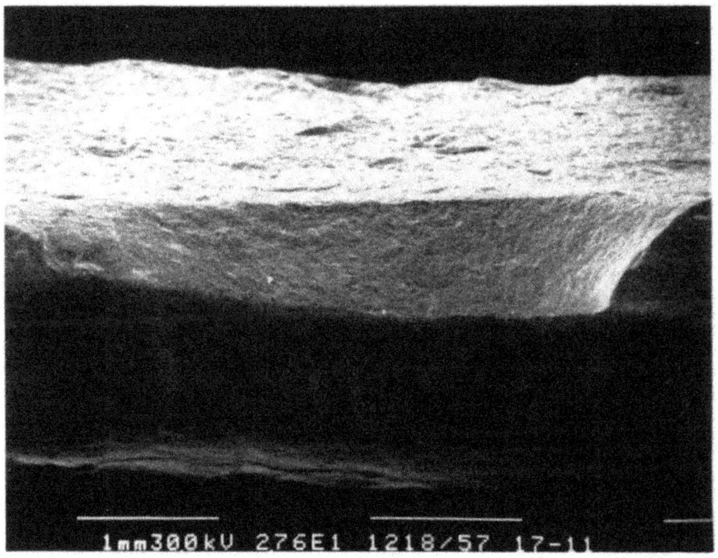

Figure 94. Very large snap-terminated scar on the edge of tool 17-11. The tool was used to saw antler (400 strokes).
(27.6 x)

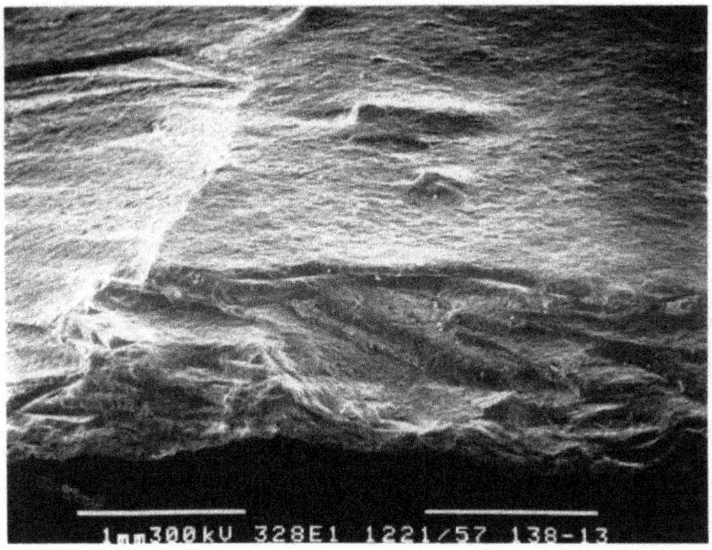

Figure 95. Step- and hinge-terminated scars on the dorsal edge-aspect of tool 138-13. The tool was used to scrape antler (115 strokes).
(32.8 x)

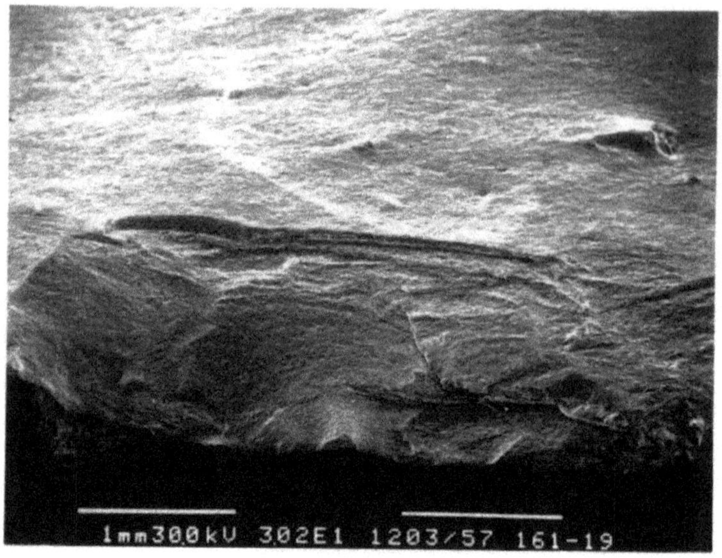

Figure 96. Large step-terminated scar with smaller scars superposed on the dorsal edge-aspect of tool 161-19. The tool was used to scrape antler (200 strokes).
(30.2 x)

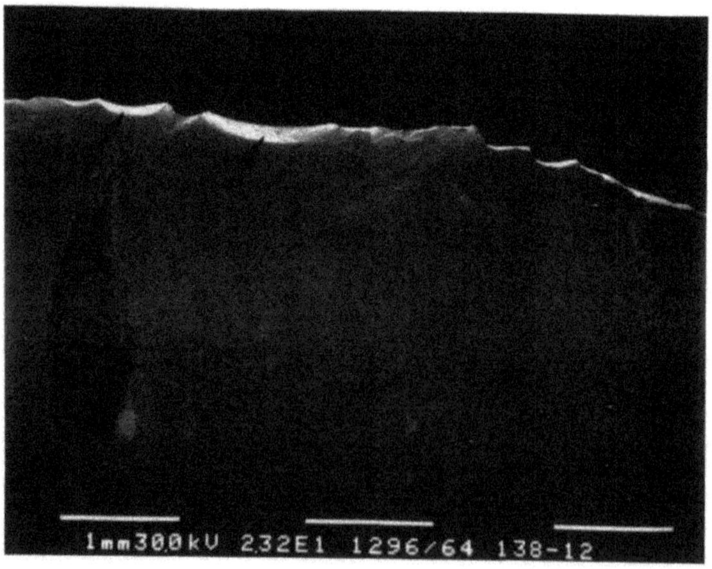

Figure 97. Small and large snap-terminated scars (arrows) on the edge of tool 138-12. The tool was used to butcher fish (1000 strokes).
(23.2 x)

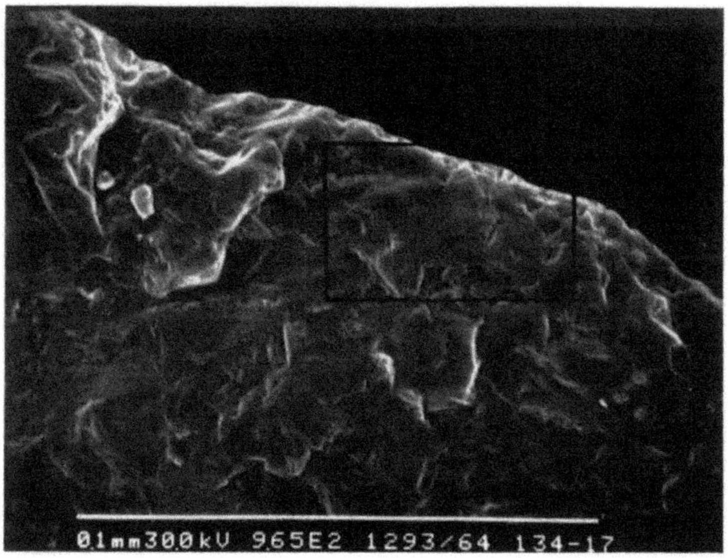

Figure 98. Smooth polish on the ventral edge-aspect of tool 134-17. The tool was used to scale fish (500 strokes).
(965 x)

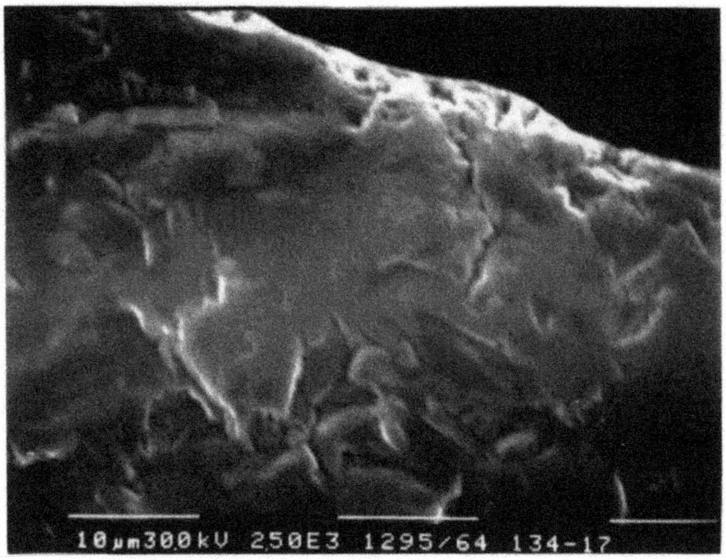

Figure 99. Close-up view of polish on tool 134-17 outlined on *Figure 98.* The polish generally smooth, although "crevices" and pock-marks are present, some of which could be described as "pits".
(2500 x)

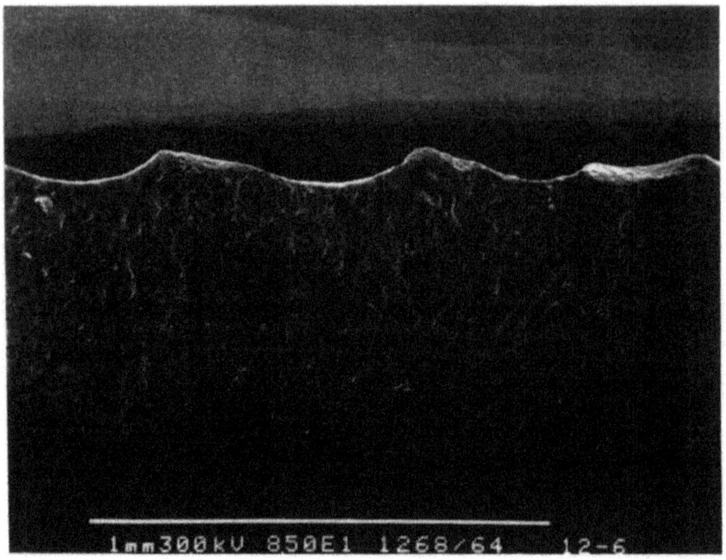

Figure 100. Snap-terminated scars and polish (arrows) on projecting parts of the denticulated edge of tool 12-6. The tool was used to skin a deer (2800 strokes).
(85 x)

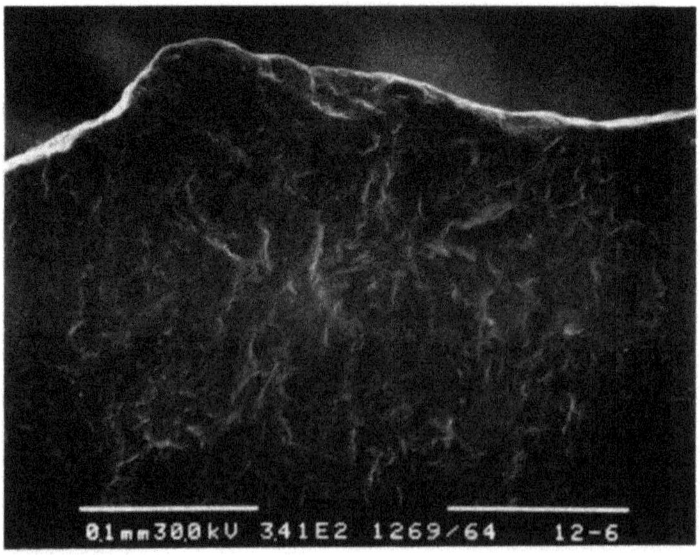

Figure 101. Close-up view of smooth polish on a projecting part of the edge of tool 12-6 shown in the center of *Figure 100.*
(341 x)

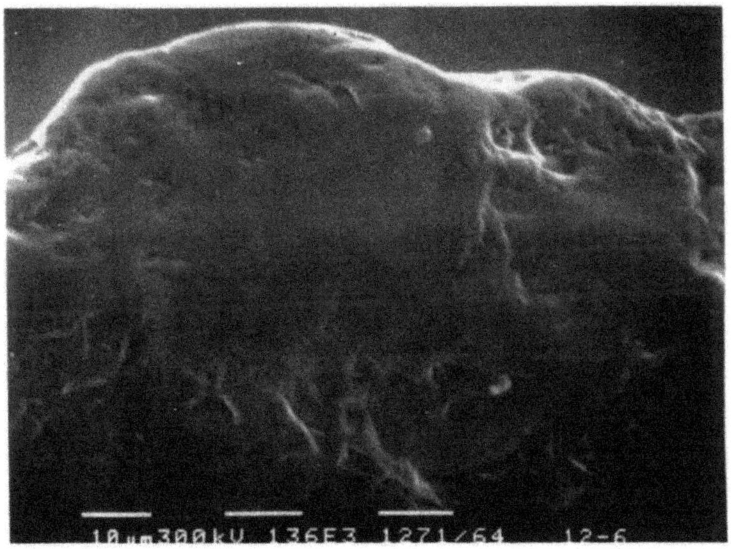

Figure 102. Close-up view of polish on tool 12-6 shown in *Figure 101.*
(1360 x)

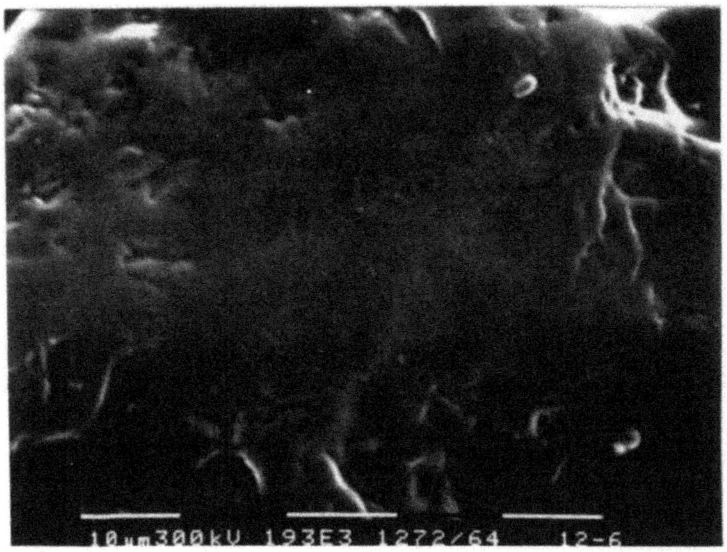

Figure 103. Close-up view of polish on tool 12-6 visible in the center-left of *Figure 102.* Faint scratches are now visible on the polish.
(1930 x)

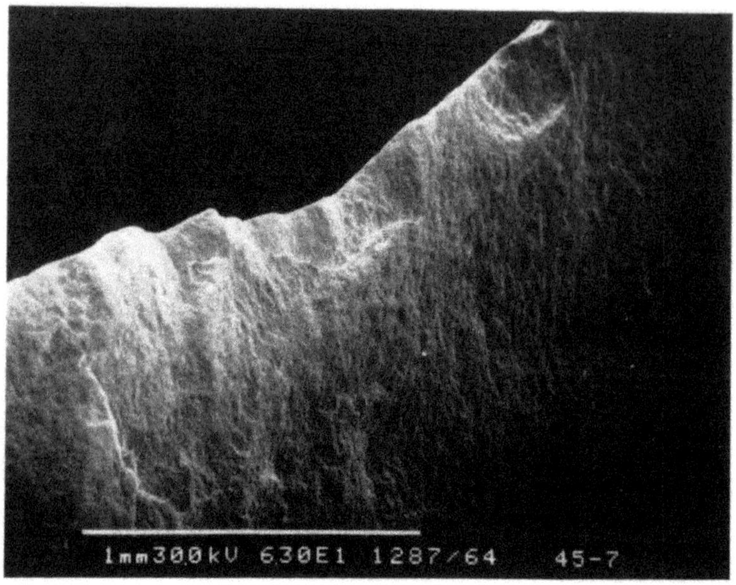

Figure 104. Oblique view along the edge and ventral edge-aspect of tool 45-7, showing shallow microflake scars. The tool was used to cut flesh (2000 strokes).
(63 x)

Figure 105. New area of tool 45-7 showing smooth polish with a tendency towards a "platy" texture (arrows) on the edge and dorsal edge-aspect.
(341 x)

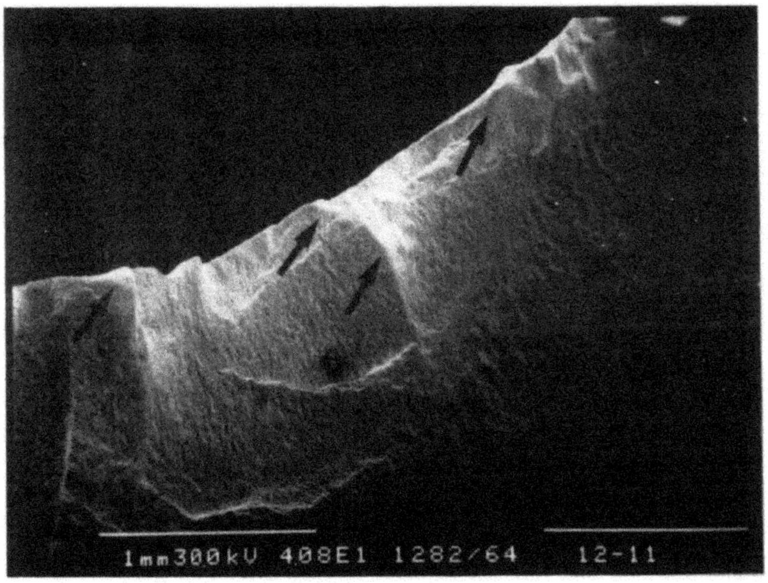

Figure 106. Oblique view along the ventral edge-aspect of tool 12-11. Note the large, shallow, step-terminated microflake scars and smoothing of scar ridges and projecting parts of the edge (arrows). The tool was used to dismember a deer (900 strokes). (40.8 x)

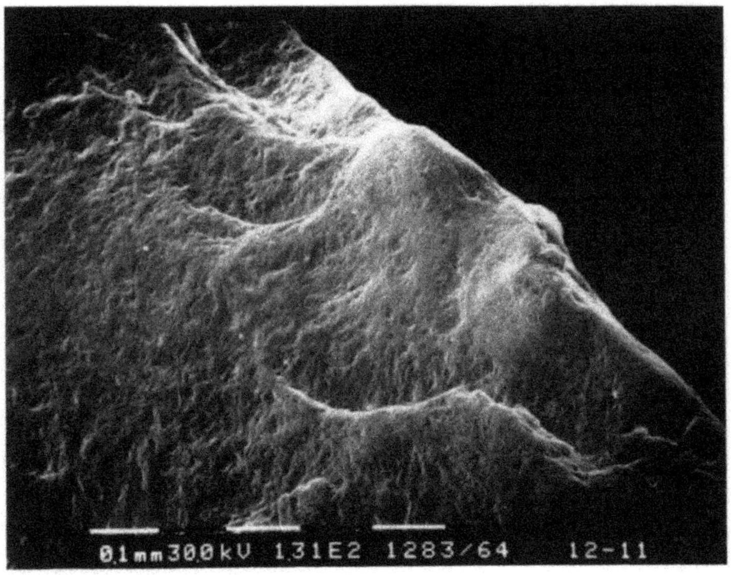

Figure 107. Smooth-pitted polish on the rounded edge and scar ridges of a new area on the ventral edge-aspect of tool 12-11. (131 x)

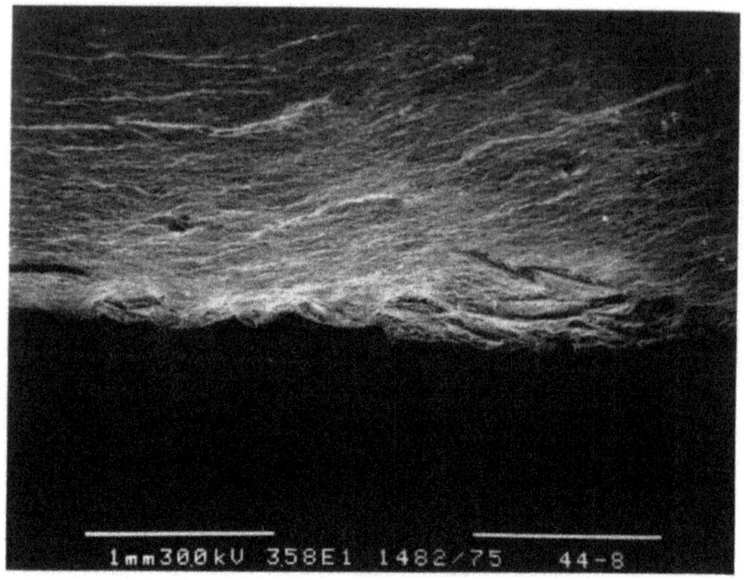

Figure 108. Microflake scars on the dorsal edge-aspect of tool 44-8. The tool was used to deflesh deer bone (1000 strokes). (35.8 x)

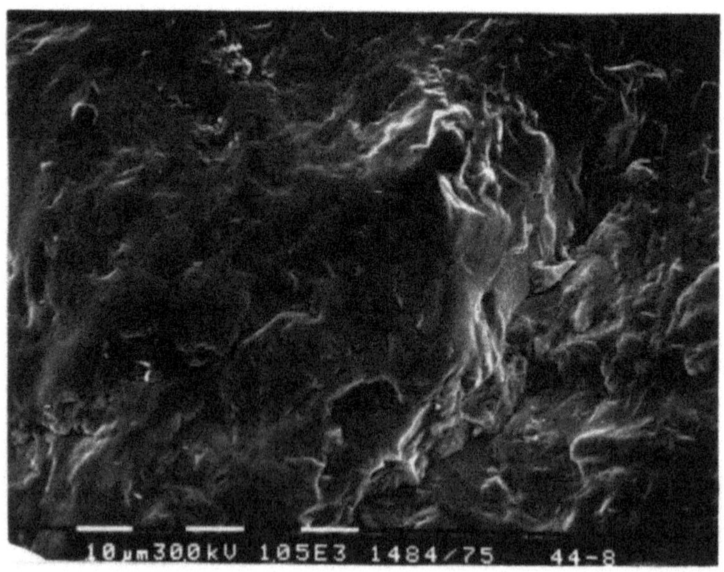

Figure 109. Smooth polish on a new part of the dorsal edge-aspect of tool 44-8.
(1050 x)

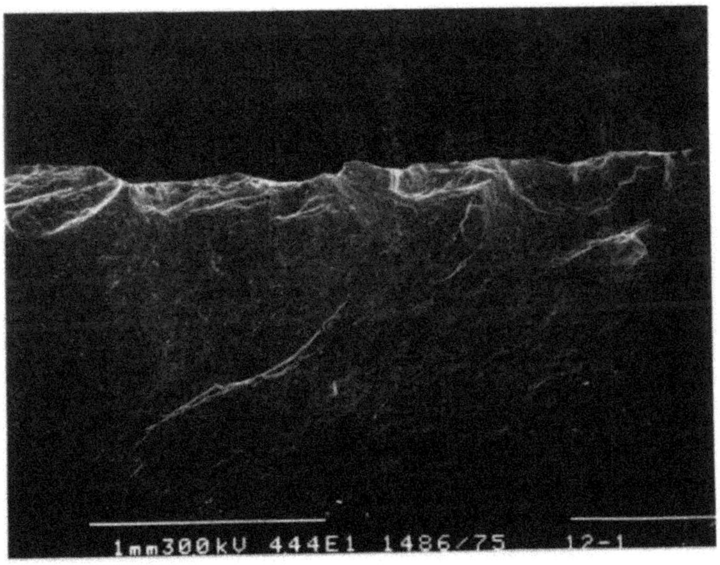

Figure 110. Microflake scars on the ventral edge-aspect of tool 12-1. The tool was used to deflesh deer bone (1000 strokes). (44.4 x)

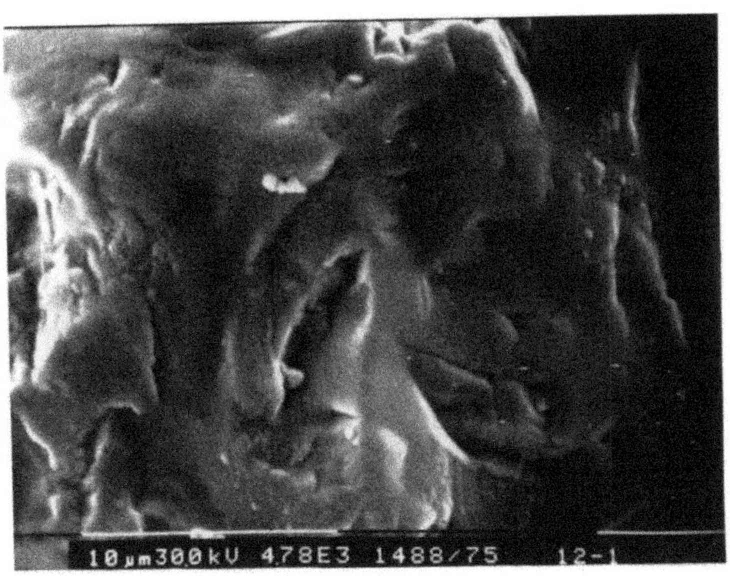

Figure 111. Smooth polish on a projection on the edge of tool 12-1. Possible pits and striations are also visible. (4780 x)

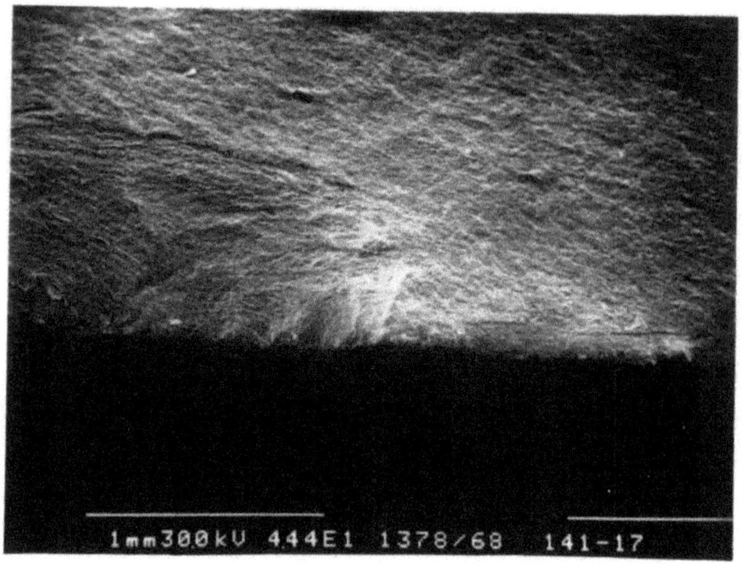

Figure 112. Step-terminated microflake scars on the dorsal edge-aspect of tool 141-6. The tool was used to scrape wood (500 strokes).
(44.4 x)

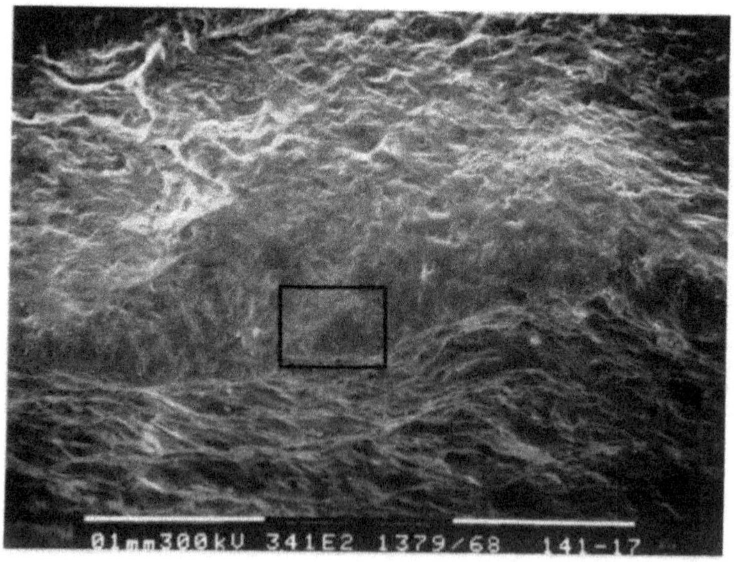

Figure 113. Extensive smooth polish on a new part of the dorsal edge-aspect of tool 141-6.
(341 x)

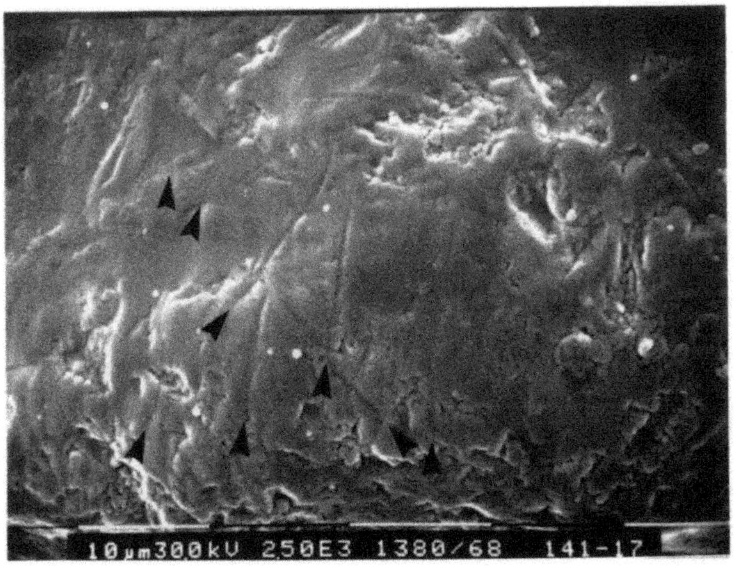

Figure 114. Close-up of area outlined on *Figure 113*. Note the primarily smooth polish broken by irregularly-shaped unpolished depression or possible pits. Numerous shallow striations (arrowheads) are also present.
(2500 x)

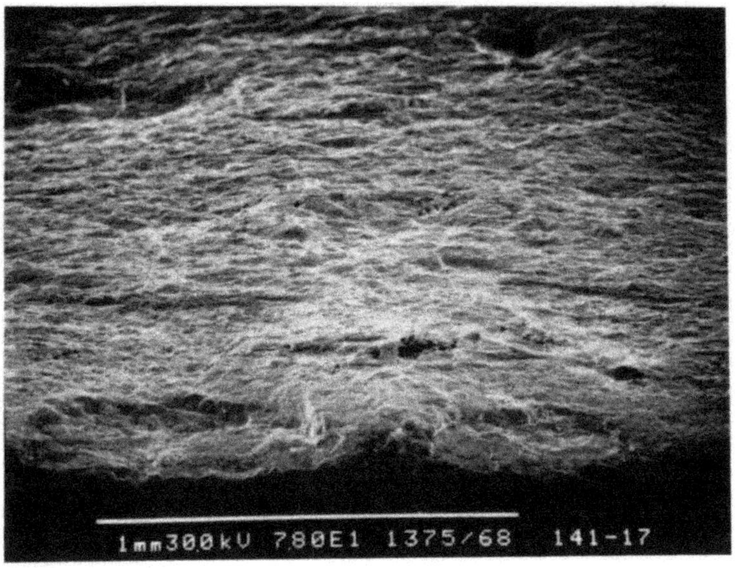

Figure 115. Microflake scars on the dorsal edge-aspect of tool 141-17. The tool was used to scrape wood (500 strokes).
(78 x)

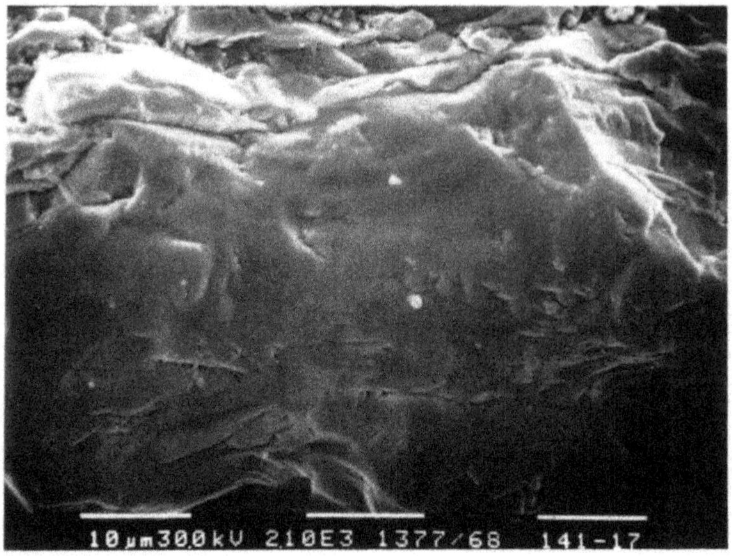

Figure 116. Smooth polish on the edge and ventral edge-aspect of a new area of tool 141-17.
(2100 x)

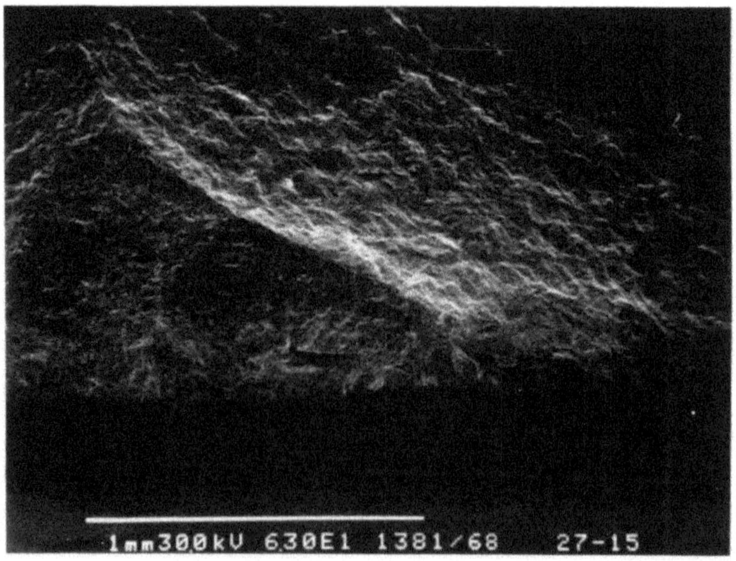

Figure 117. Microflake scars on the dorsal edge-aspect of tool 27-15. The tool was used to scrape wood (2000 strokes).
(63 x)

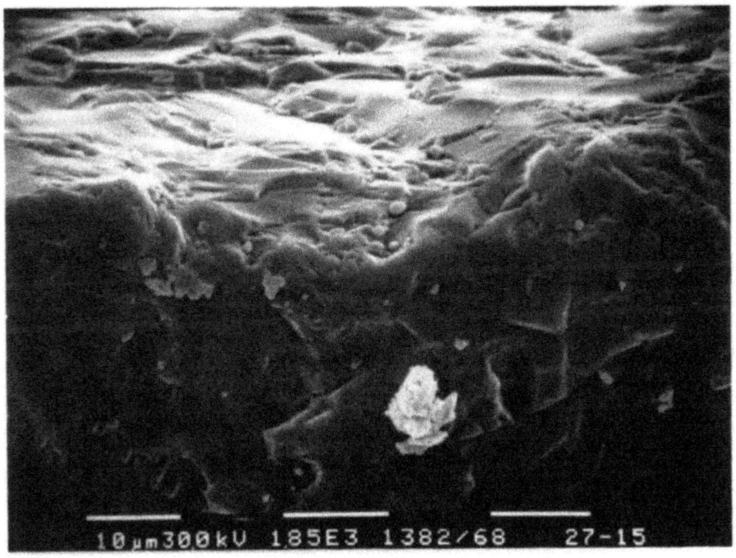

Figure 118. Smooth polish and a striation (arrowhead) on the edge and ventral edge-aspect of tool 27-15.
(1850 x)

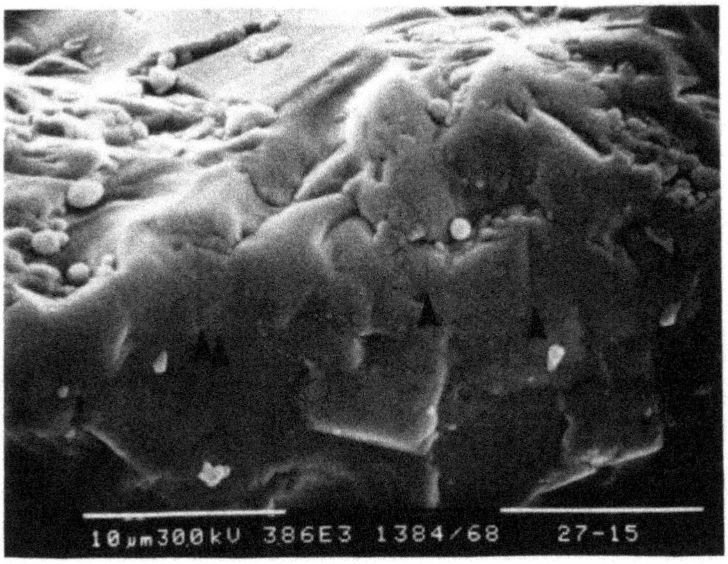

Figure 119. Close-up view of polished area with striation on tool 27-15 shown to the center-right of *Figure 118.* Several more striations (arrowheads) and possible striations are now visible.
(3860 x)

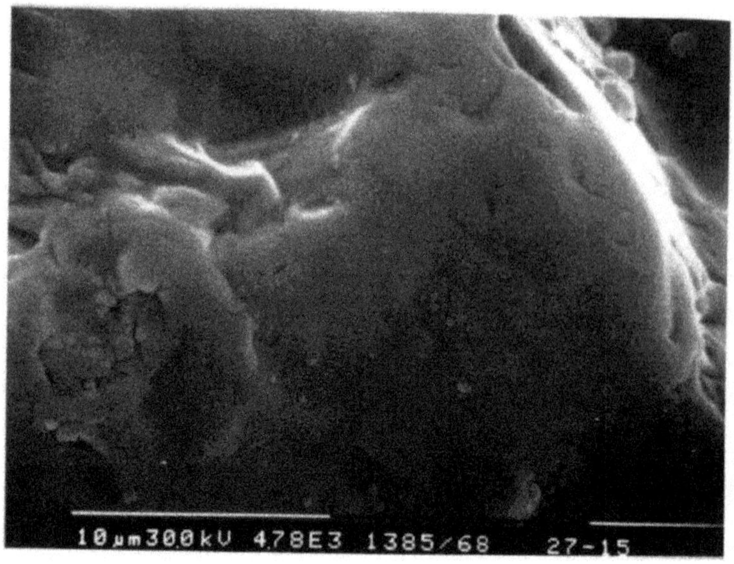

Figure 120. Smooth polished surface on a new part of the ventral edge-aspect of tool 27-15.
(4780 x)

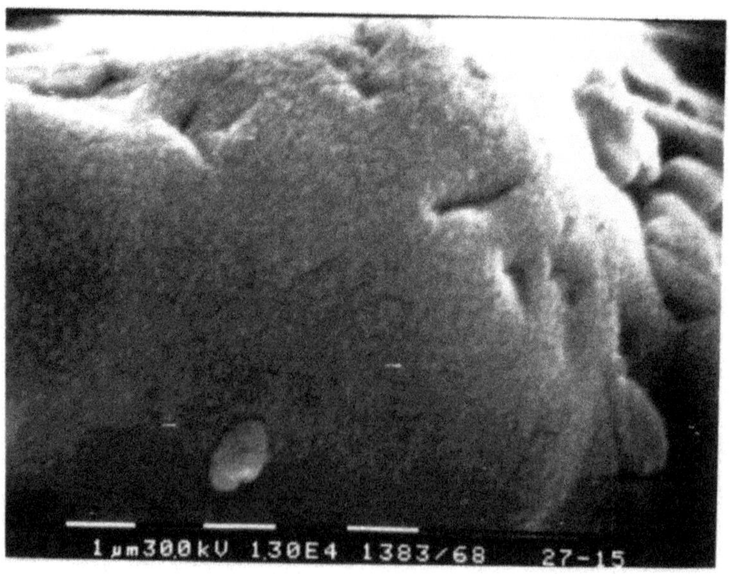

Figure 121. Extreme close-up view of the polish on tool 27-15 shown in *Figure 120.* The polish still appears smooth, but with a slightly "pebbley" texture at this magnification.
(13,000 x)

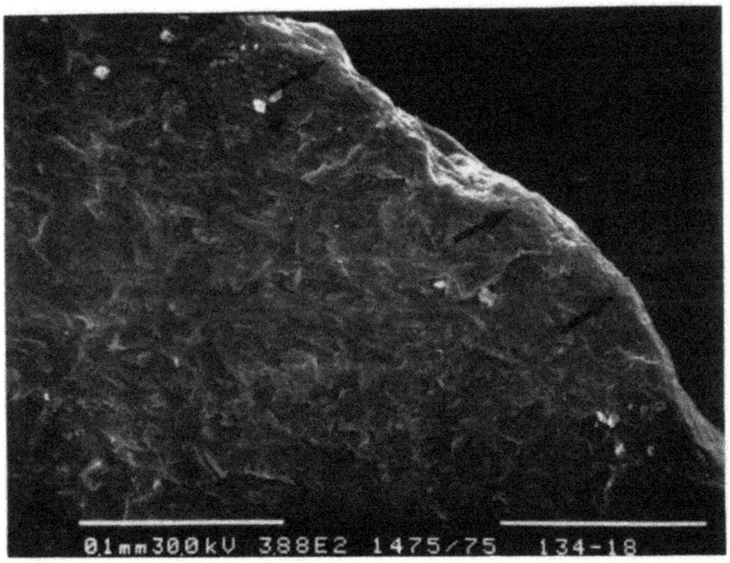

Figure 122. Smooth-pitted polish (arrows) and rounding visible on the edge and ventral edge-aspect of tool 134-18. The tool was used to saw wood (1000 strokes).
(388 x)

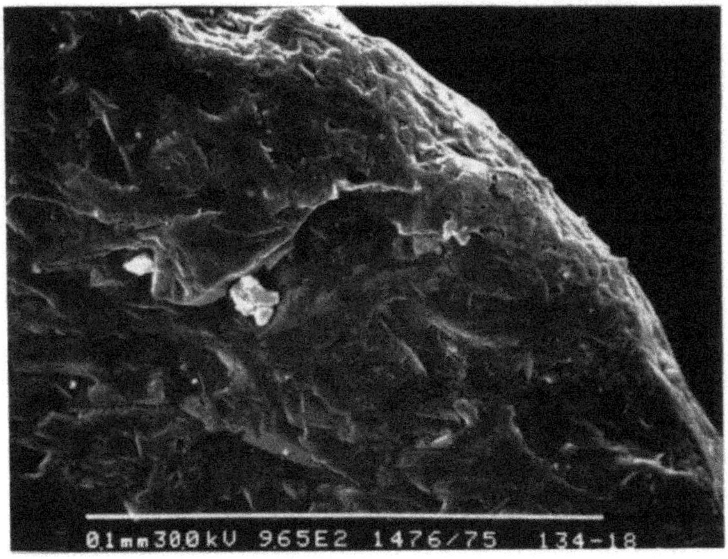

Figure 123. Close-up view of the polish on tool 134-18 shown in the center-right of *Figure 122.* Note the smooth polish with a somewhat pitted or "leafy" appearance.
(965 x)

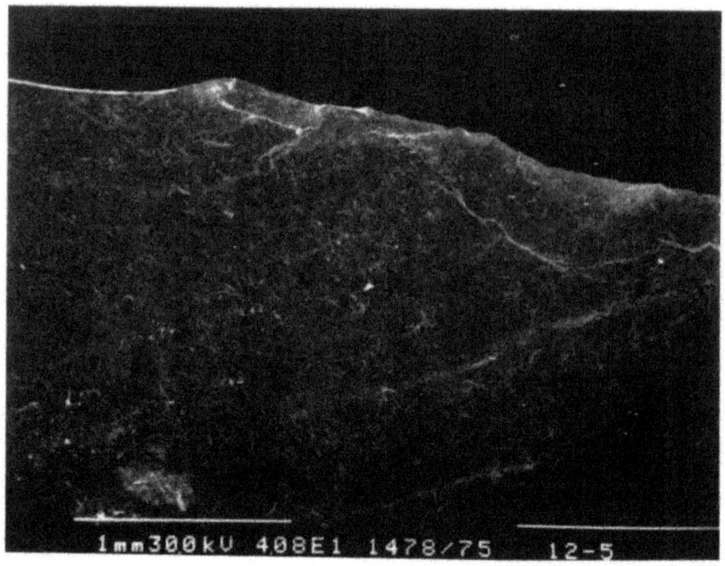

Figure 124. Microflake scars on the ventral edge-aspect of tool 12-5. The tool was used to whittle wood (500 strokes). (40.8 x)

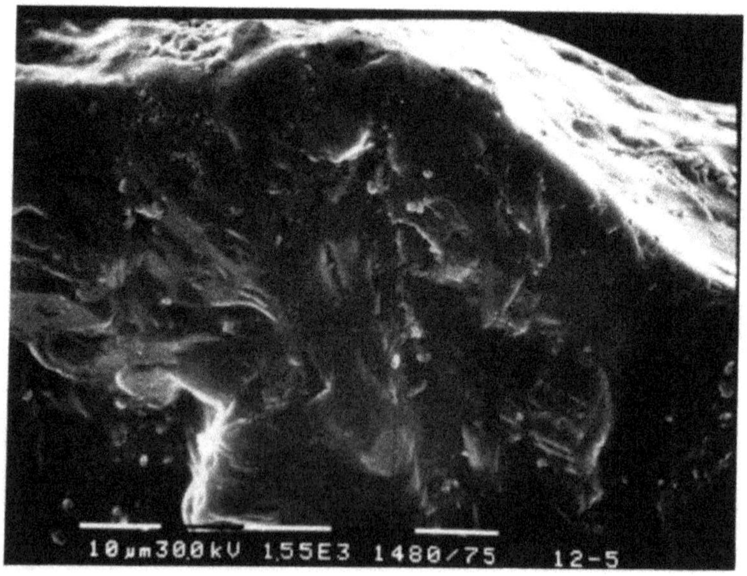

Figure 125. Smooth polish on the slightly rounded edge and dorsal edge-aspect of a new part of tool 12-5. (1550 x)

APPENDIX D

OUTLINE DRAWINGS OF EXPERIMENTAL TOOLS

Legend for Tool Illustrations

Cortex: 

Used Part of Tool: 

Extent of Polish on Used Edge: 

Note: Each tool is oriented with its platform to the bottom of the page and its dorsal face up.

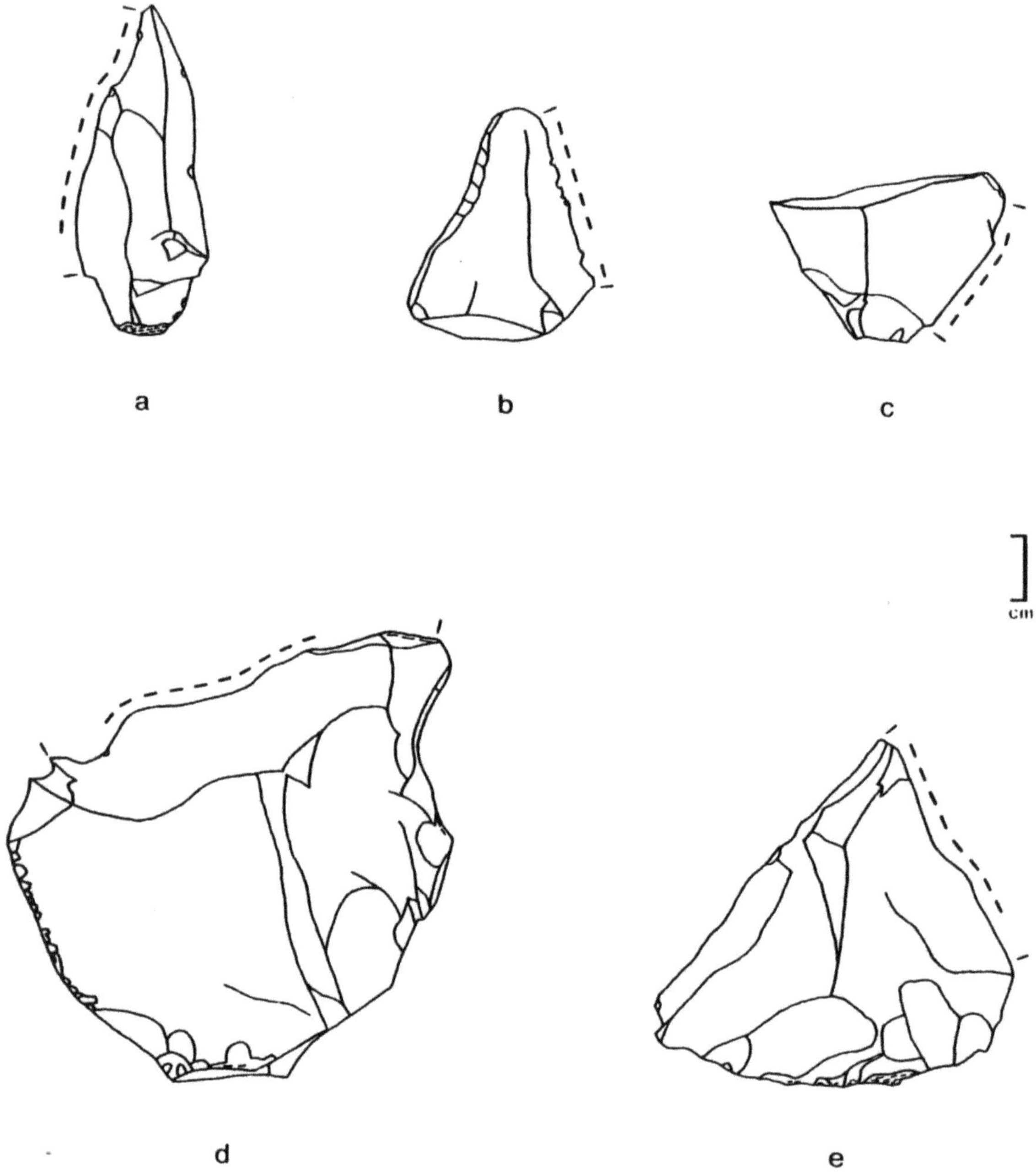

Figure 126. Plant cutting tools. a: 12-9, b: 161-11, c: 43-2, d: 134-11, e: 12-12.

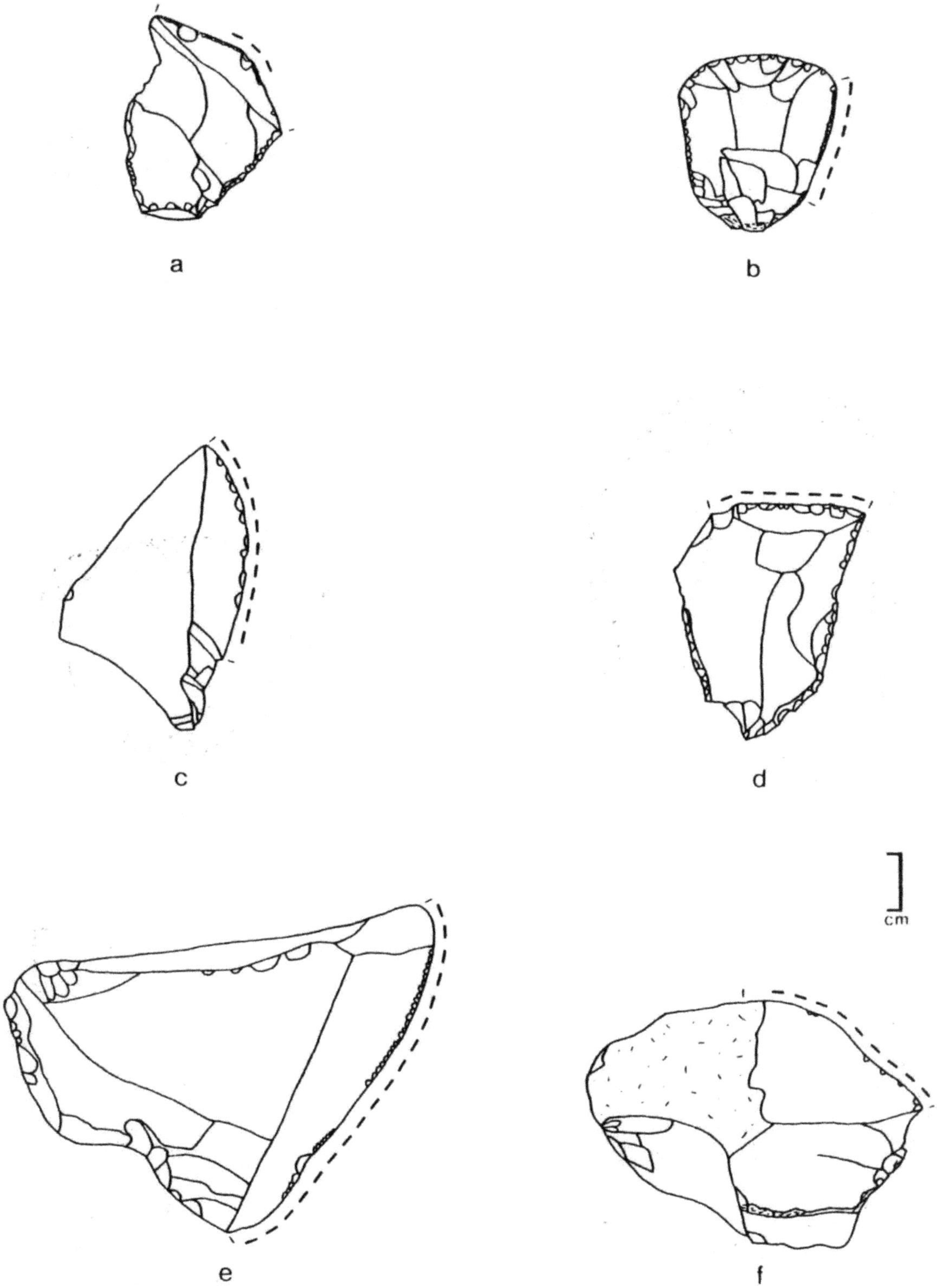

Figure 127. Hide working tools. a: 44-4, b: 22-7, c: 23-3, d: 12-13, e: 17-14, f: 27-16.

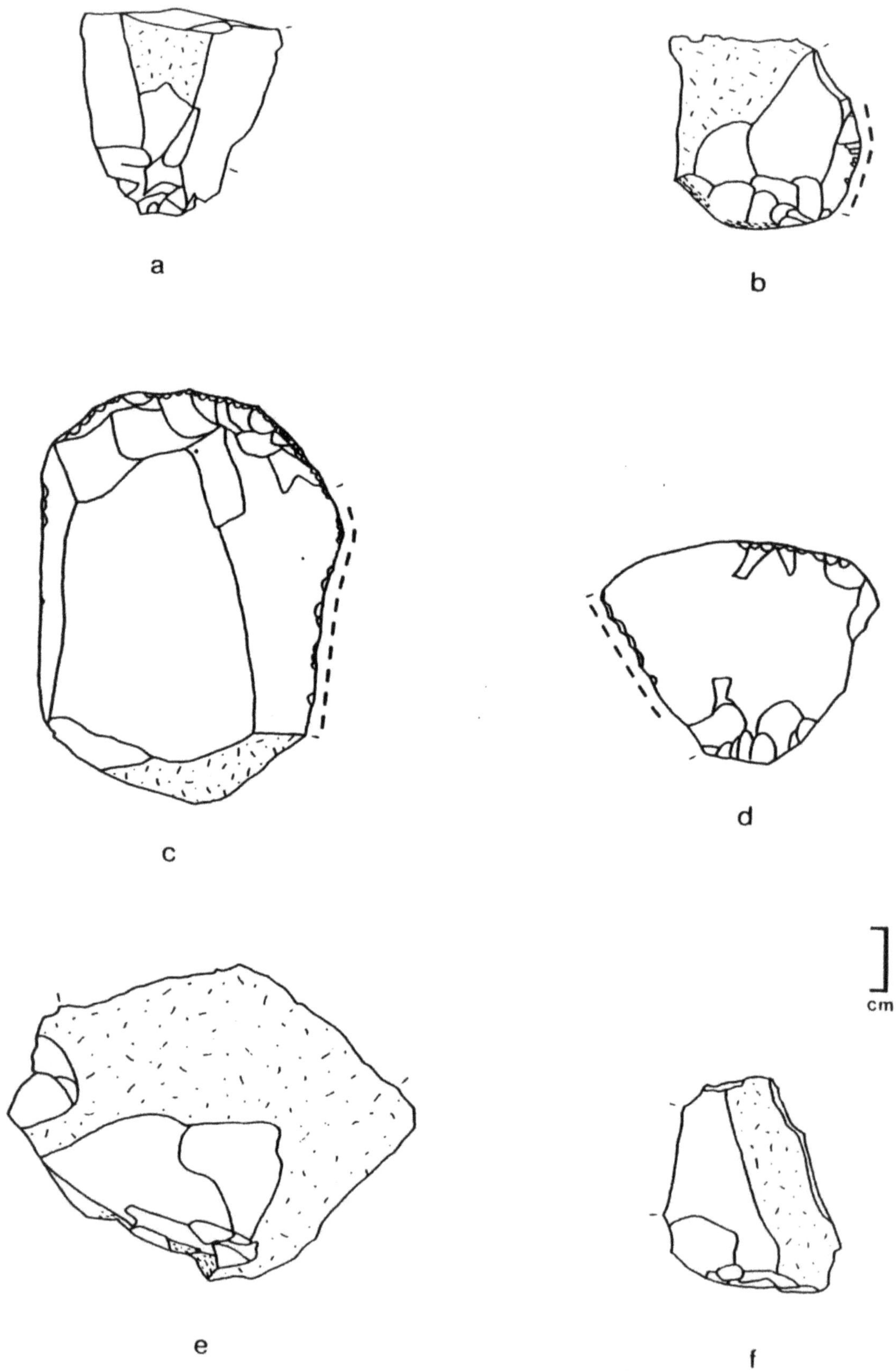

Figure **128**. Hide working tools. a: 1-3, b: 39-1, c: 134-9, d: 12-17, e: 80-6, f: 3-4.

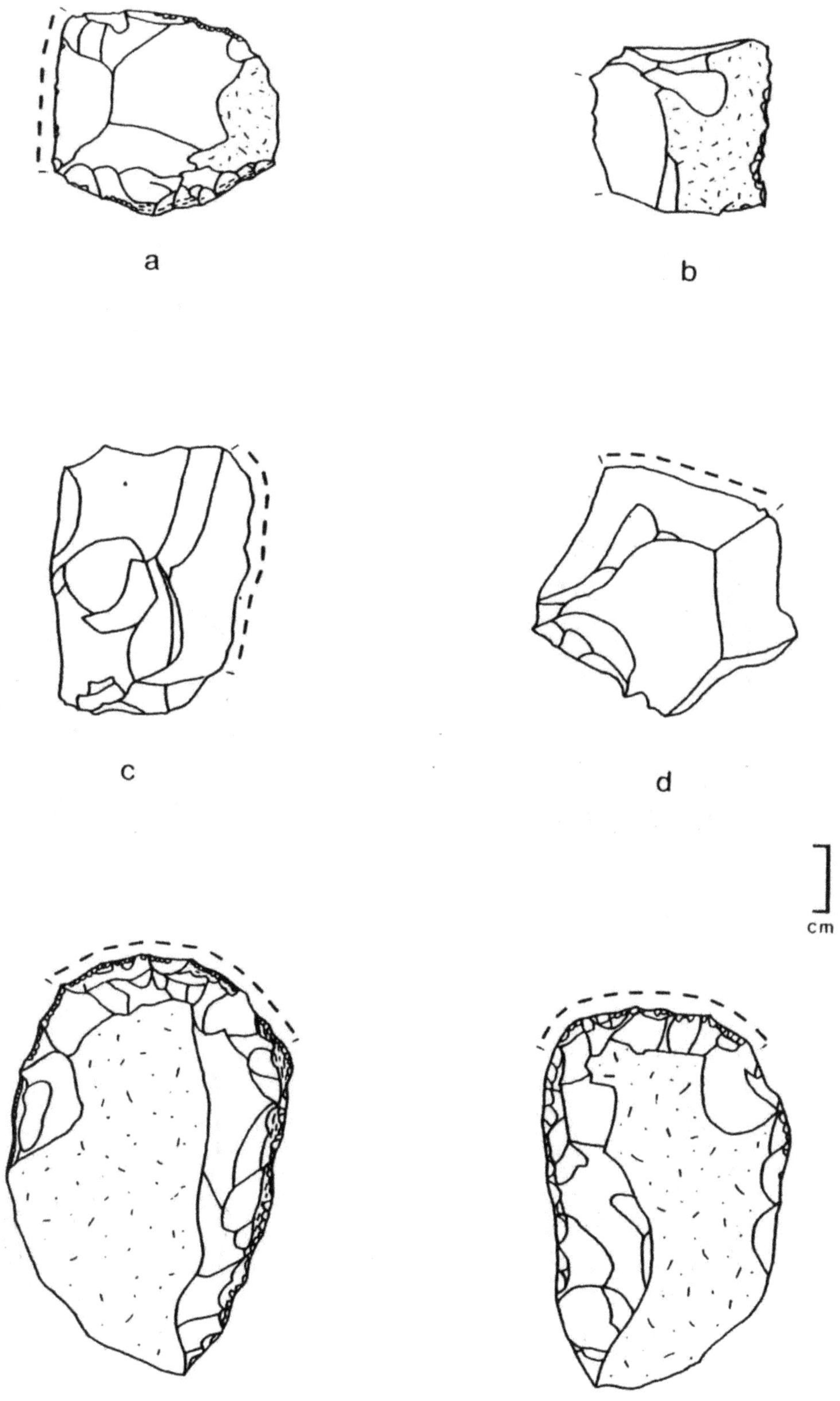

Figure 129. Hide working tools. a: 46-4, b: 31-2, c: 80-7, d: 80-8, e: 80-4, f: 80-5.

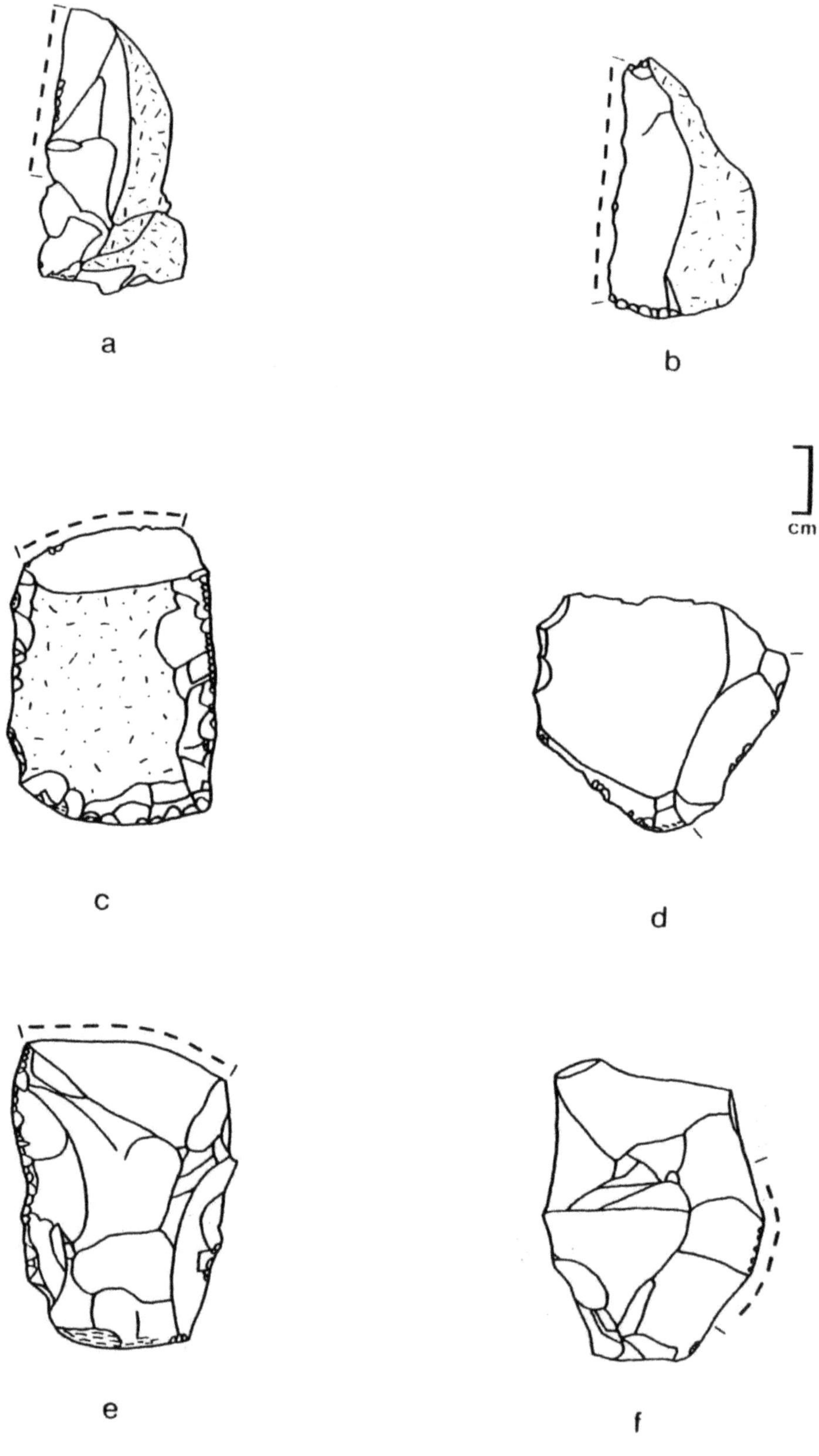

Figure 130. Bone working tools. a: 22-20, b: 21-6. Hide working tools. c: 22-12, d: 3-3, e: 27-17, f: 44-6.

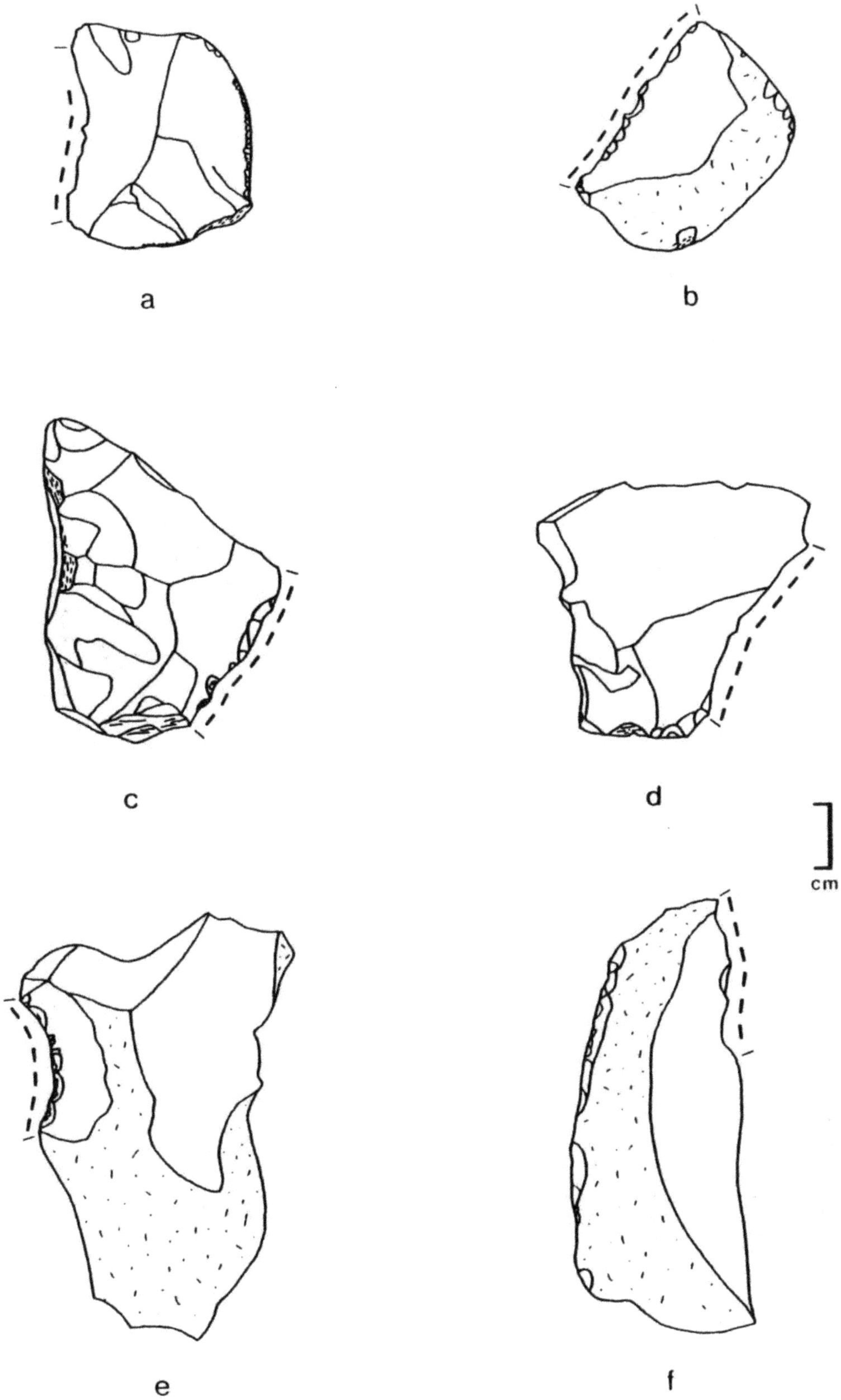

Figure 131. Bone working tools. a: 11-6, b: 46-2, c: 63-1, d: 26-9, e: 17-13, f: 80-19.

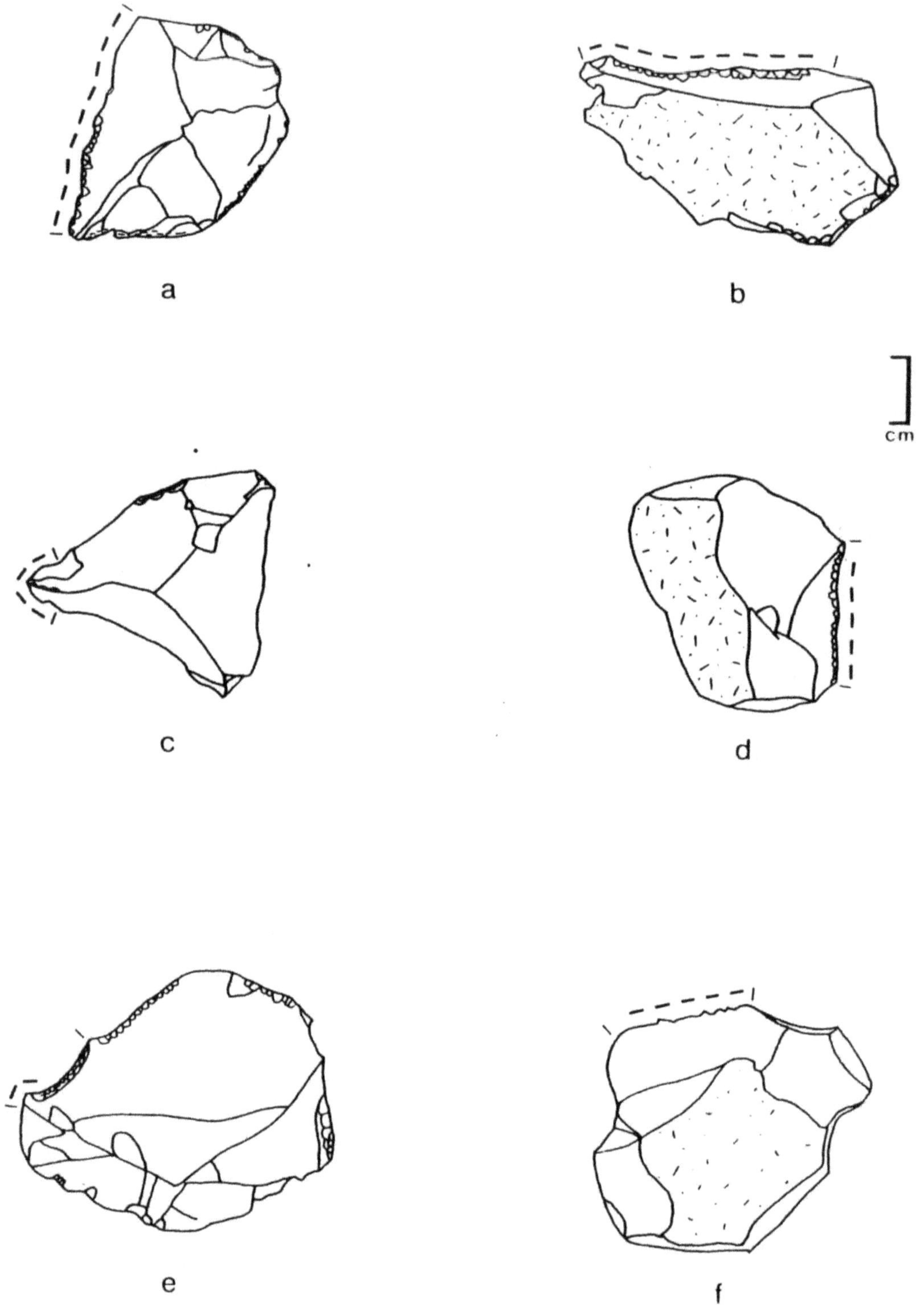

Figure 132. Antler working tools. a: 43-3, b: 22-11, c: 134-6, d: 141-14. Bone working tools. e: 138-19, f: 26-8.

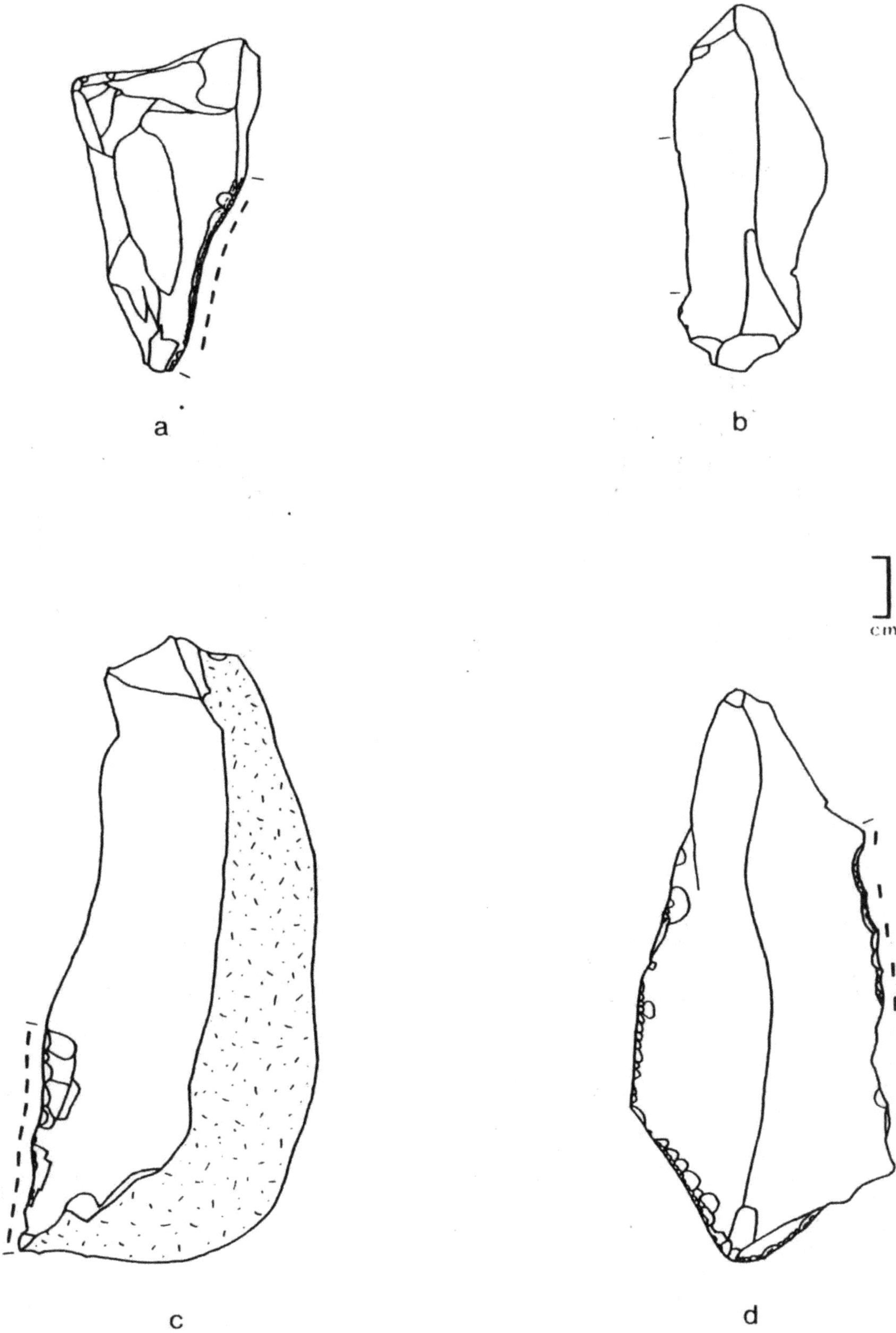

Figure 133. Antler working tools. a: 138-13, b: 11-12, c: 161-18, d: 134-22.

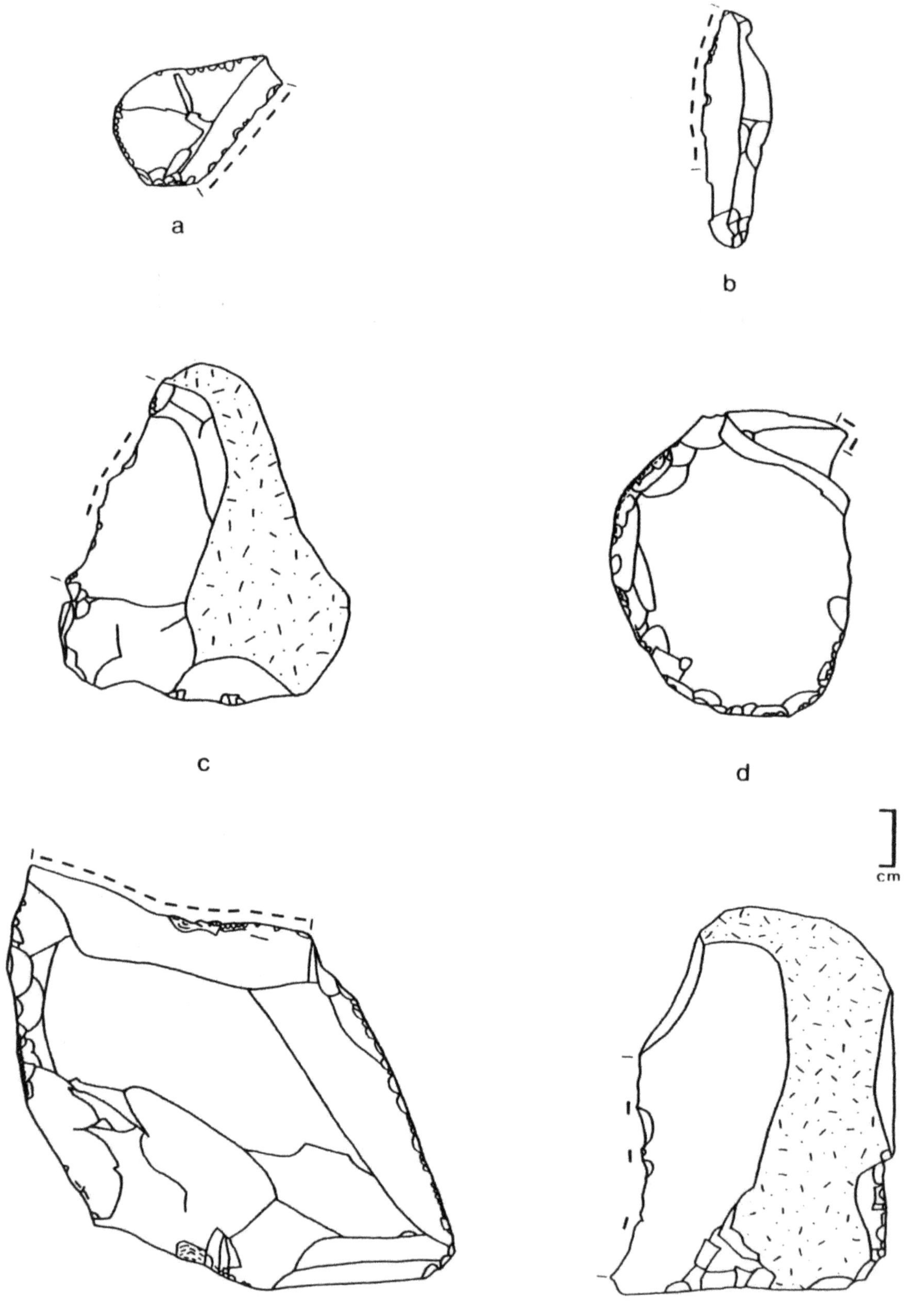

Figure 134. Antler working tools. a: 27-3, b: 36-7, c: 141-12, d: 141-20, e: 161-19, f: 11-12.

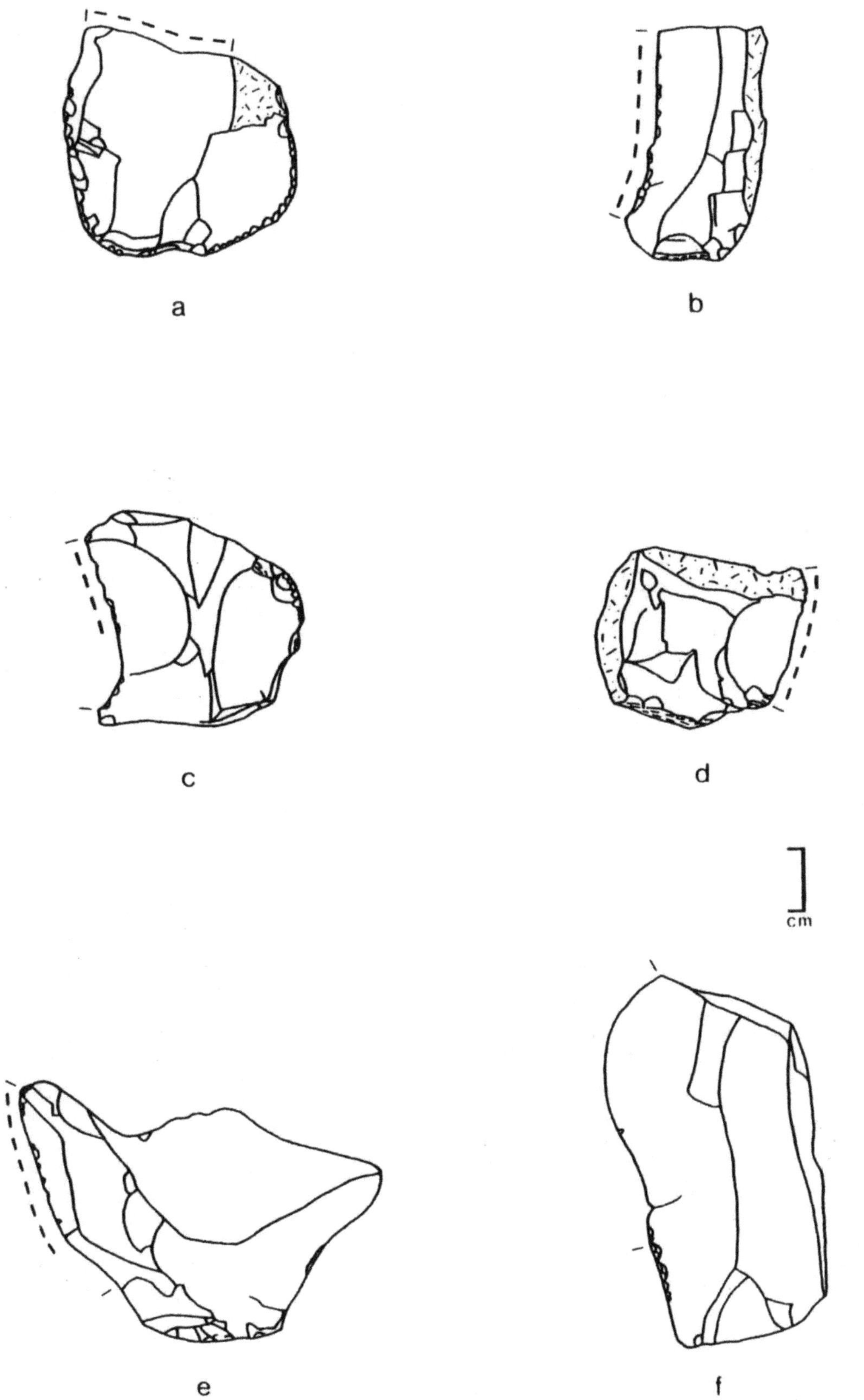

Figure 135. Antler working tools. a: 17-11, b: 141-7, c: 80-17.

Fish butchering tools. d: 46-3, e: 134-16, f: 138-12.

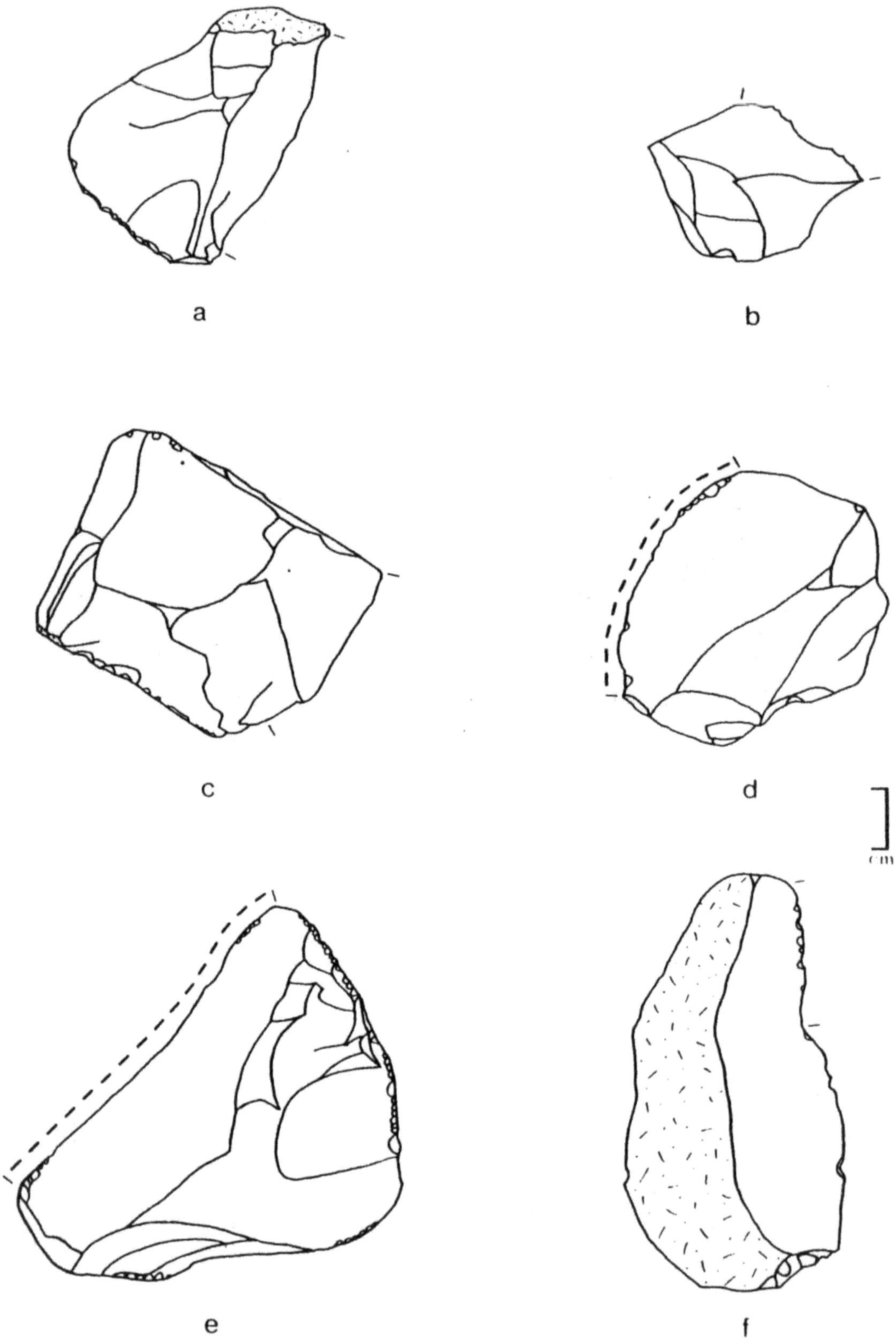

Figure 136. Fish butchering tools. a: 20-7, b: 12-14, c: 134-17, d: 12-10. Deer butchering tools. e: 12-6, f: 22-15.

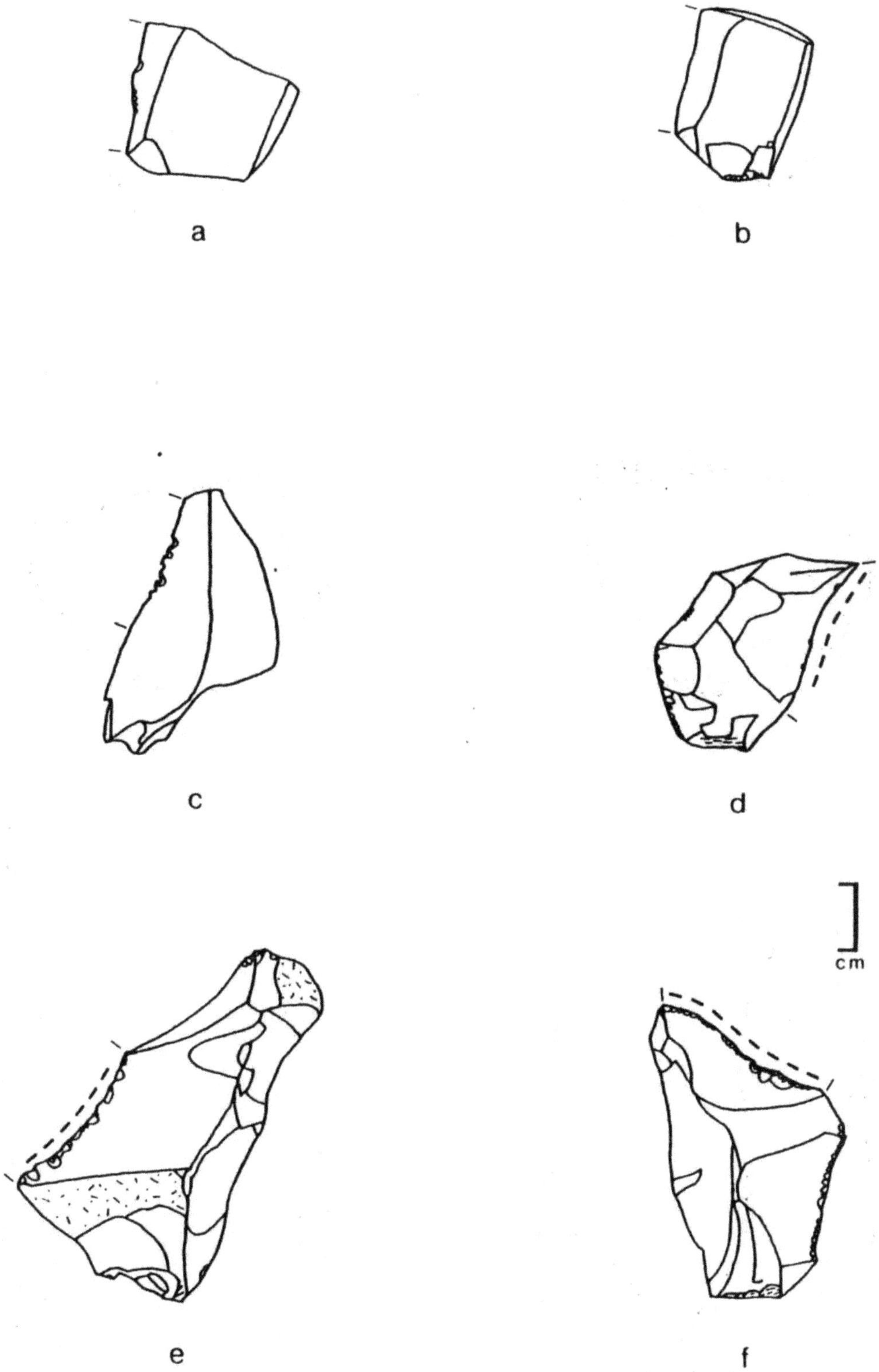

Figure 137. Deer butchering tools. a: 2-3, b: 2-1, c: 11-3, d: 12-1, e: 44-11, f: 44-8.

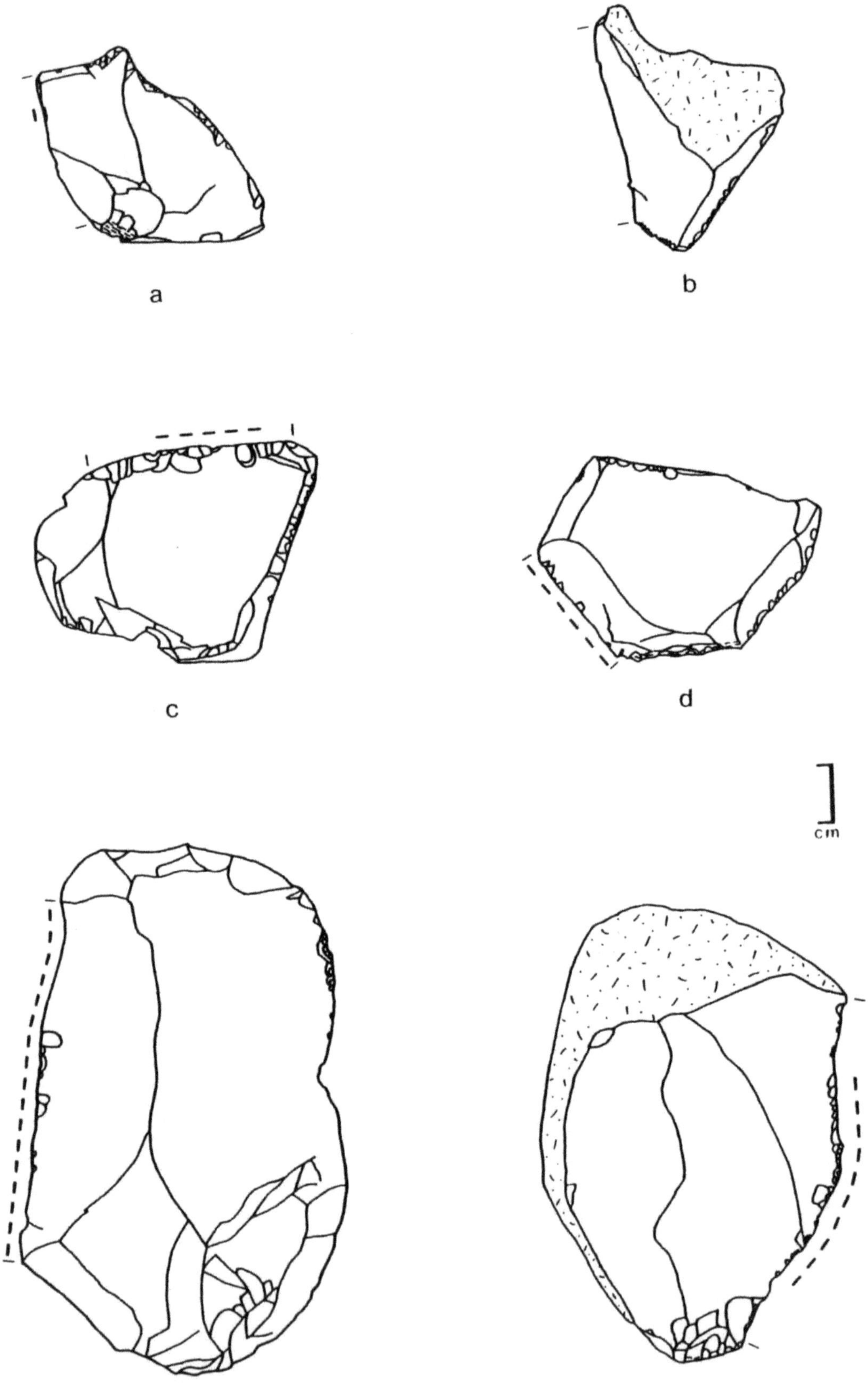

Figure 138. Deer butchering tools. a: 12-15, b: 22-10, c: 80-14, d: 45-6, e: 12-11, f: 80-3.

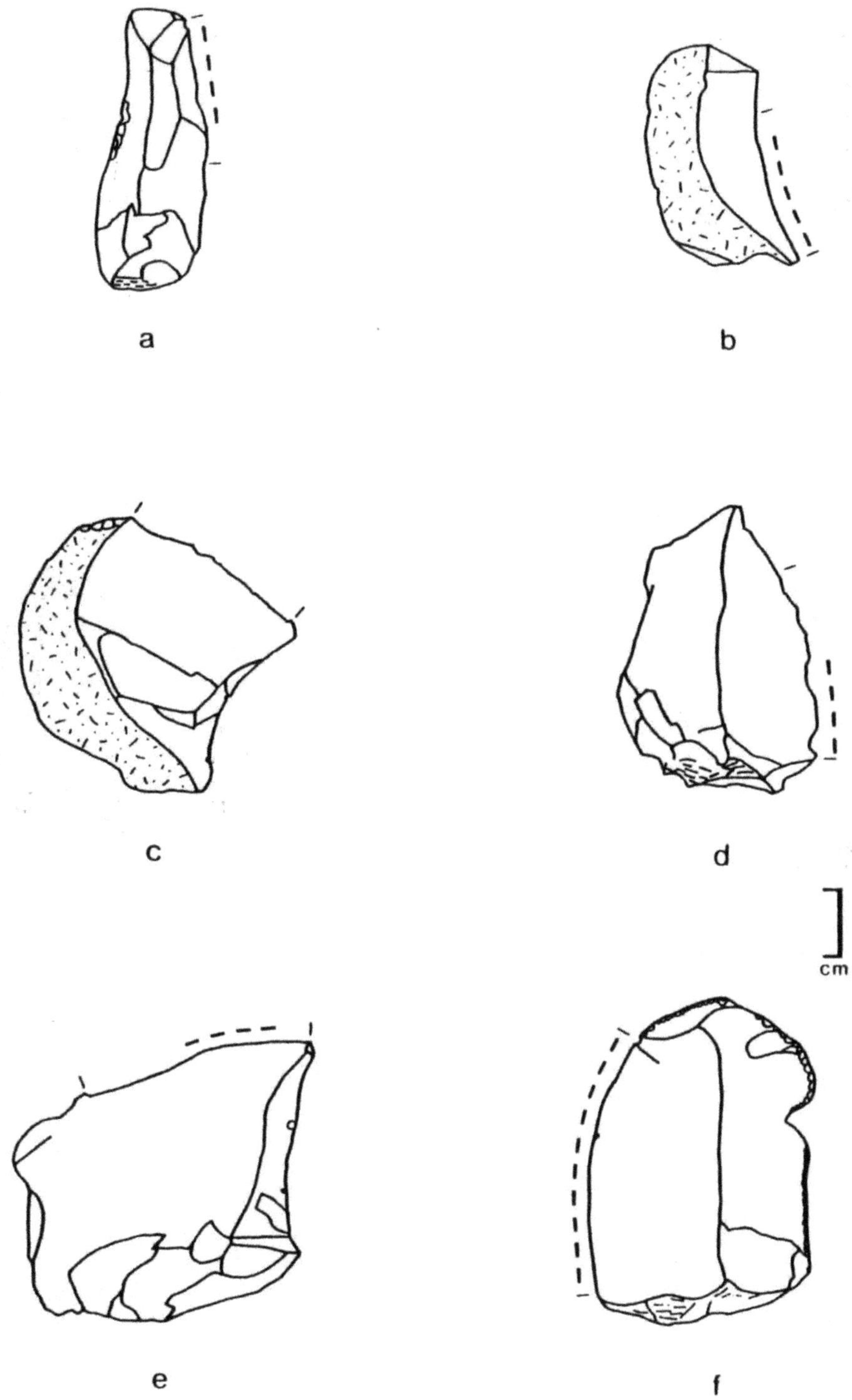

Figure 139. Deer butchering tools. a: 36-8, b: 27-5, c: 80-13, d: 1-4, e: 11-9, f: 43-4.

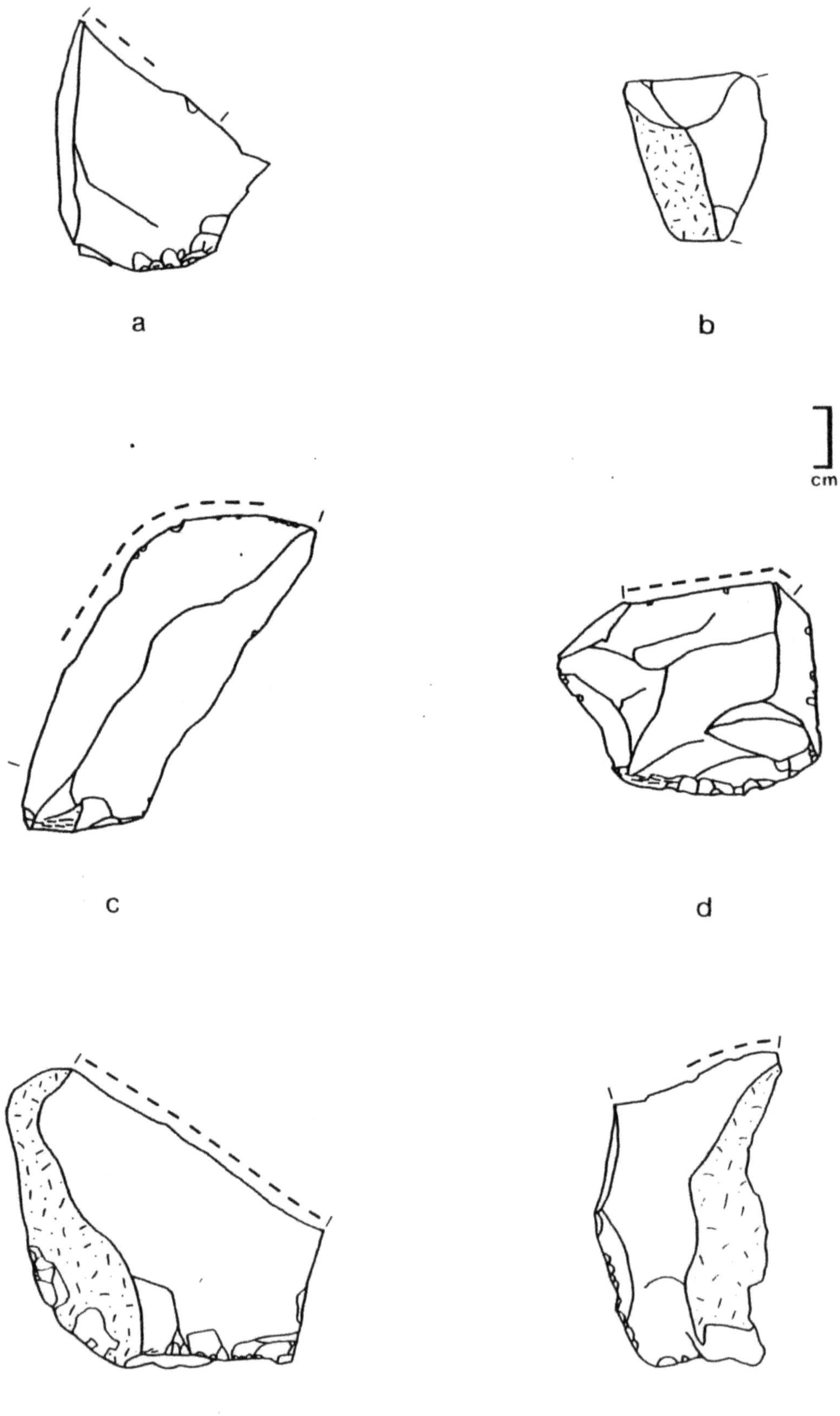

Figure **140.** Deer butchering tools. a: 22-17, b: 35-3, c: 12-3, d: 45-7, e: 80-12. Wood working tool. f: 23-4.

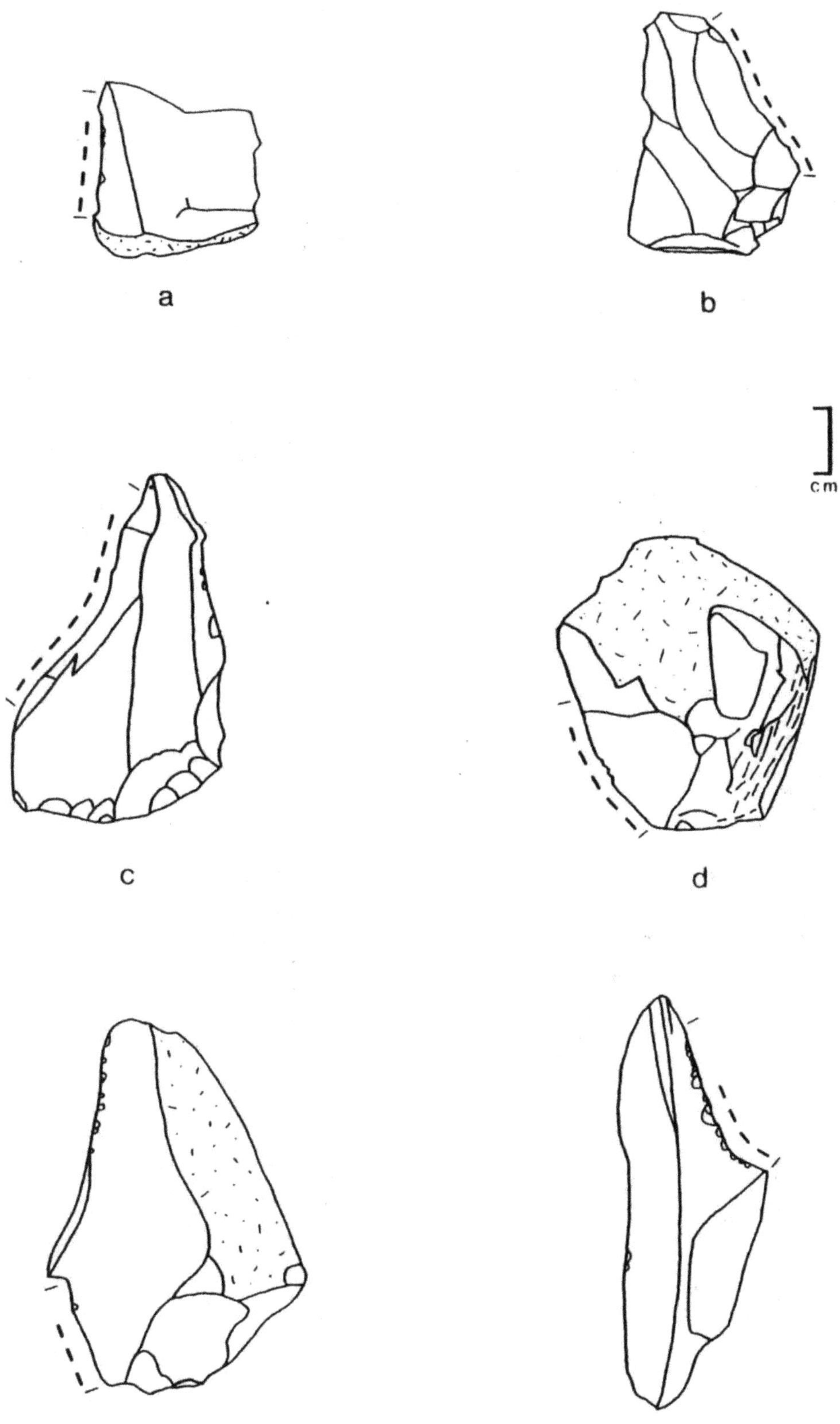

Figure 141. Wood working tools. a: 80-11, b: 80-9, c: 27-15, d: 44-10, e: 17-12, f: 138-16.

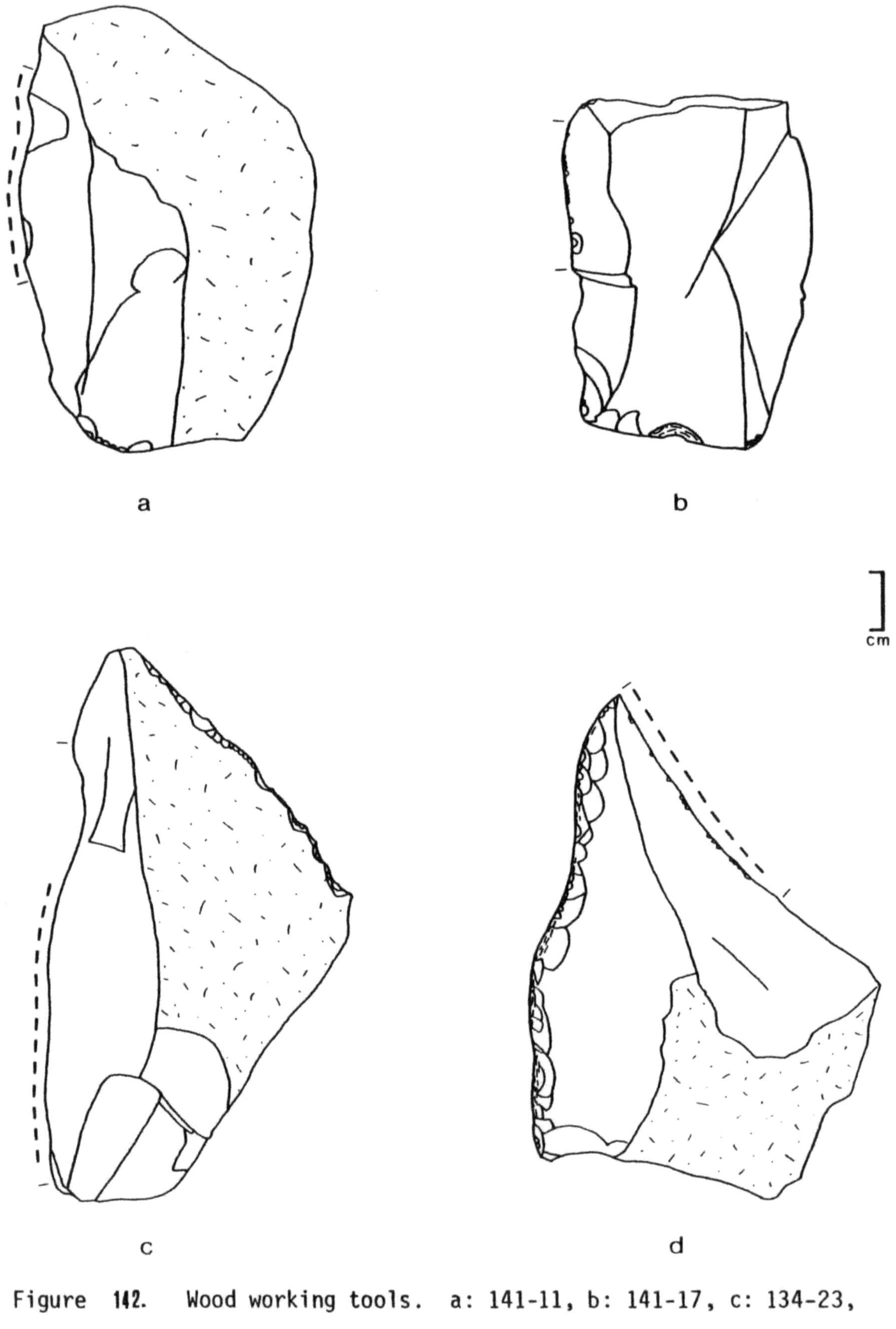

Figure 142. Wood working tools. a: 141-11, b: 141-17, c: 134-23, d: 161-20.

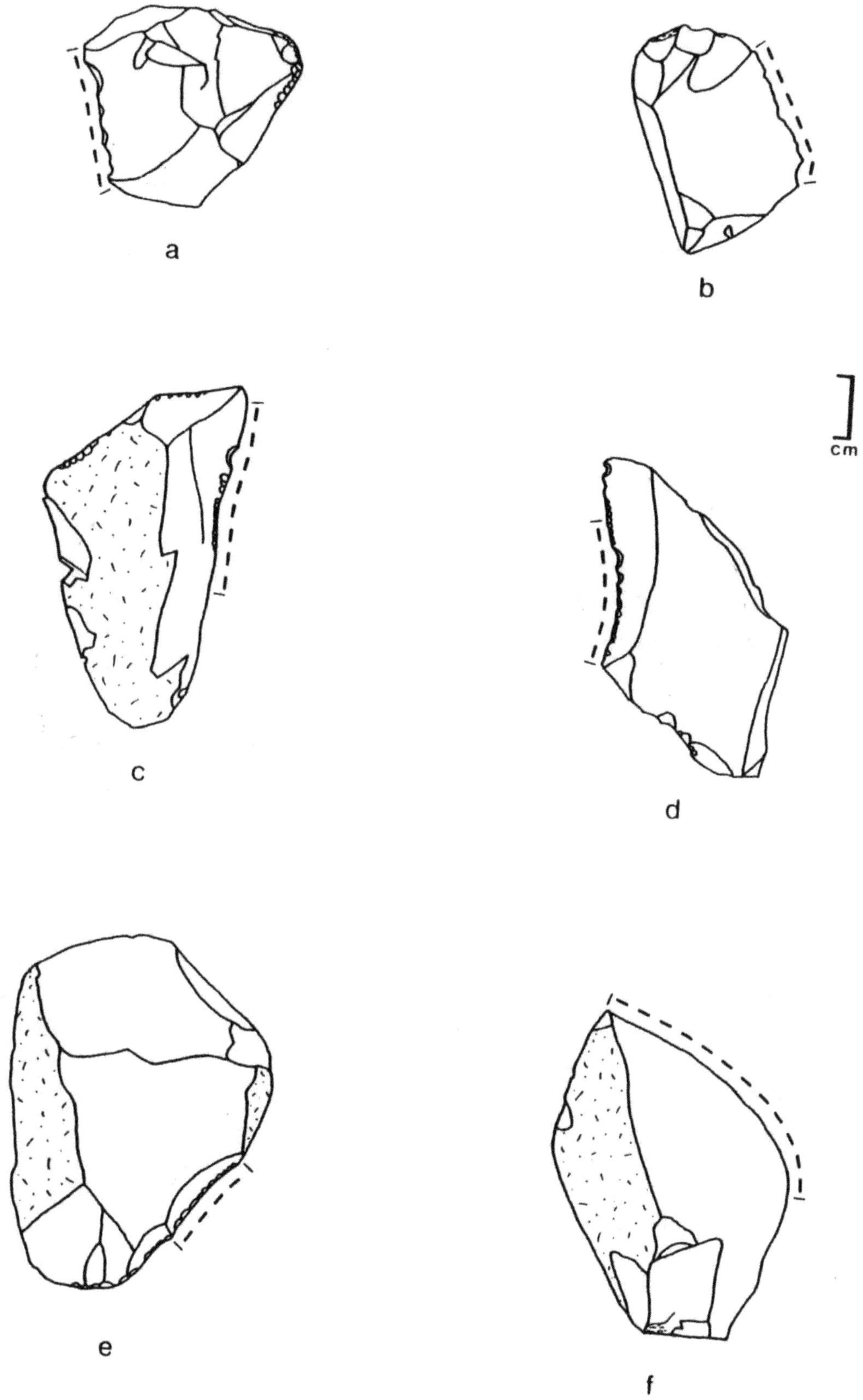

Figure 143. Wood working tools. a: 27-10, b: 27-7, c: 141-6, d: 138-6, e: 26-10, f: 44-7.

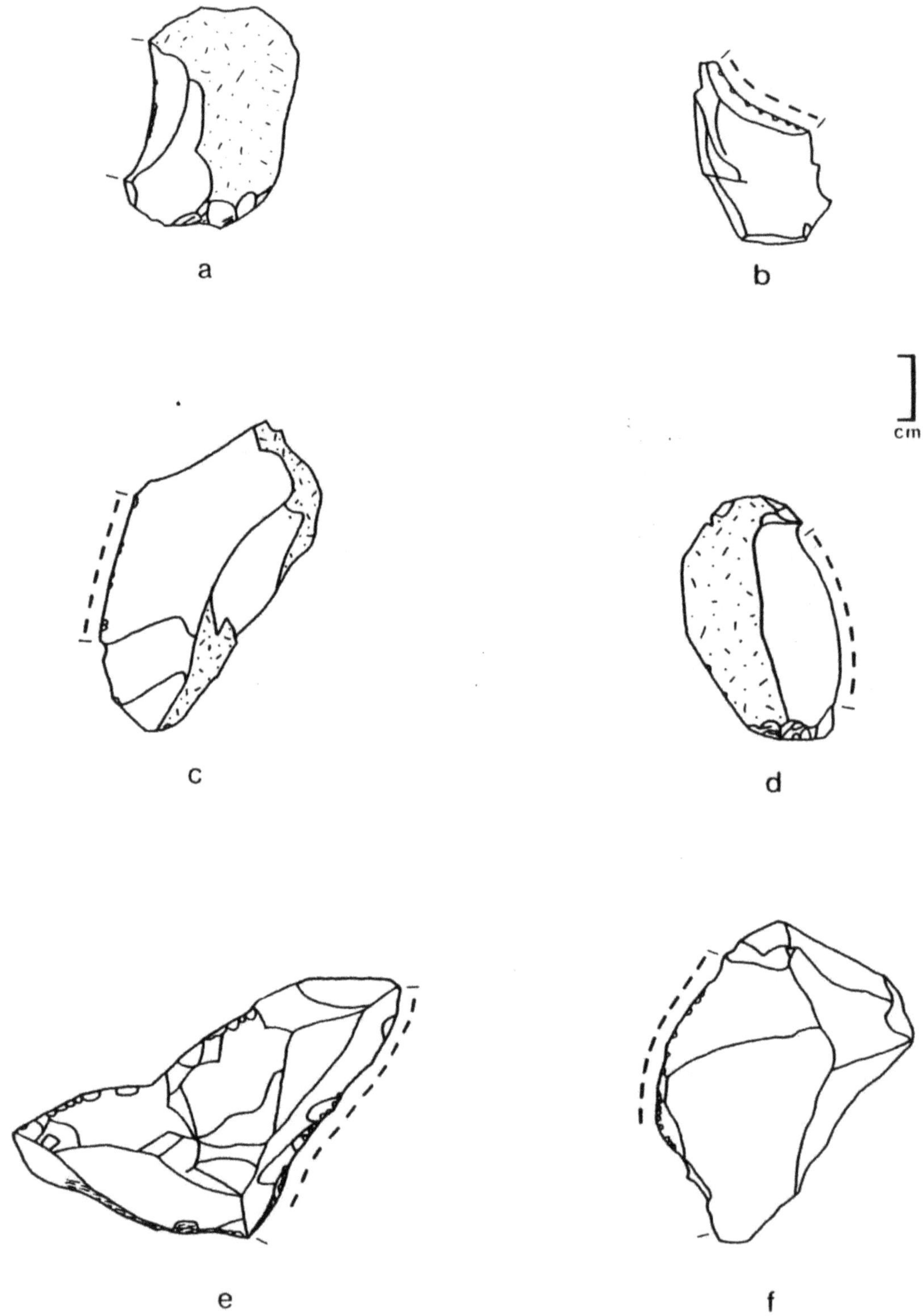

Figure 144. Wood working tools. a: 141-4, b: 80-10, c: 44-9, d: 27-8, e: 161-15, f: 27-14.

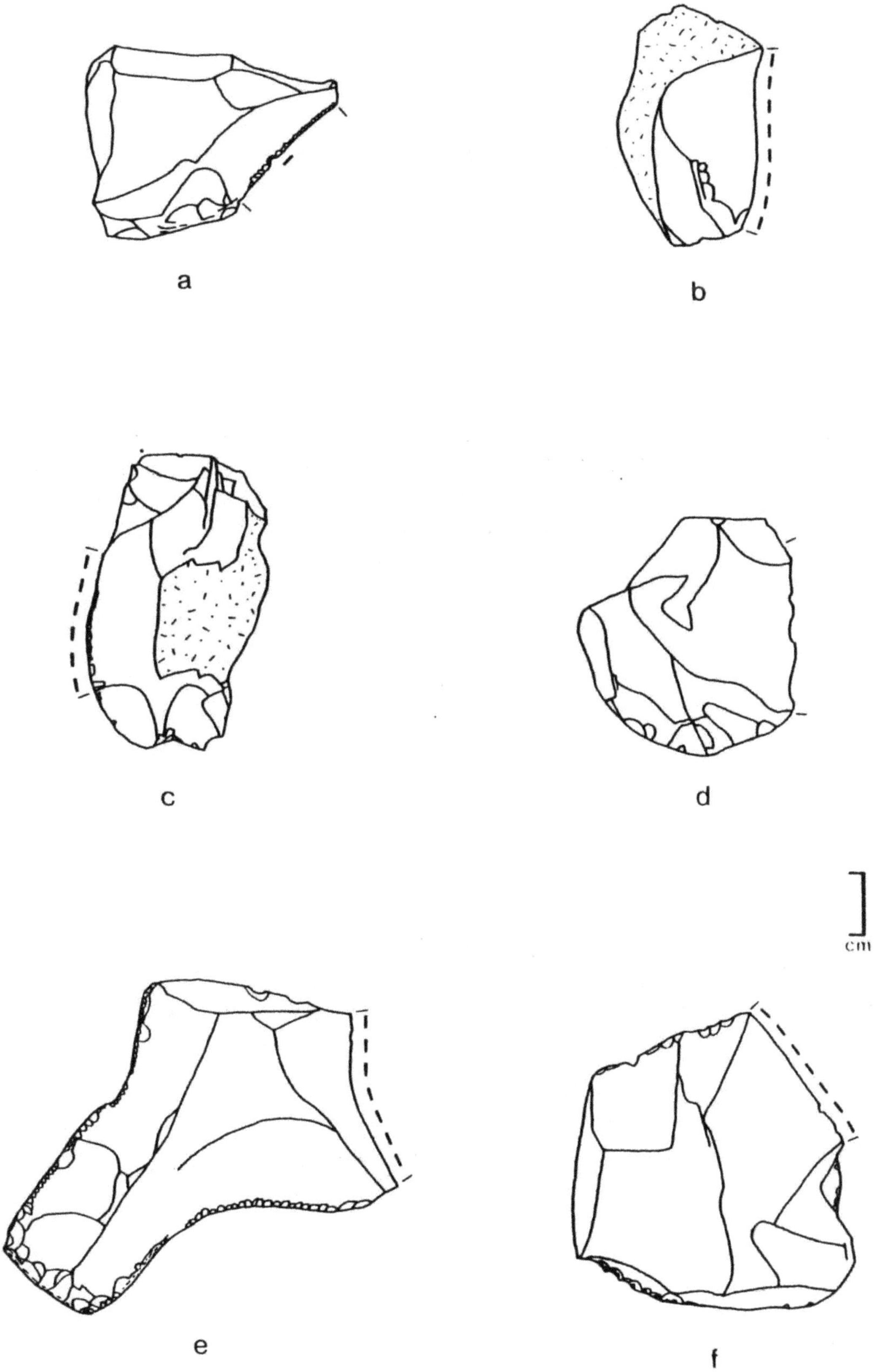

Figure 145. Wood working tools. a: 27-11, b: 82-3, c: 35-5, d: 63-2, e: 12-5, f: 134-18.

APPENDIX E

BLIND TEST INTERPRETATIONS

Introduction

This appendix contains the interpretations of stone tool function made during the three blind tests discussed in Chapter 9. The actual tool actions and contact materials are presented on Table 20. The methods used, dates the tests were undertaken and the time required for interpretation are summarized for each test. *Time* includes the number of minutes required for studying each tool, making observations, recording interpretations and indicating the used edge on a sketch of the tool. For interpretations under *Action*, the general use-action is followed by the more specific action in brackets. *Comments* generally consist of discussions of interpretations of tool action and contact material, as well as comparisons with wear patterns on experimental tools, descriptions of the wear traces observed, and uncertainties in interpretation. The comments are deliberately brief for the non-microscopic *Blind Test 1*.

Blind Test 1

Interpretations were made on the basis of tool morphology, location and characteristics of flake scars, spine-plane angle of the used edge, general knowledge of stone tool use, and intuition. A 5 x hand lens was the only aid used for observing the tools.

Dates Undertaken: June 22-23, 1986

Time: total = 140 min.; mean = 14 min.; shortest time = 10 min, longest time = 27 min.

Tool 1

Used area(s): 1

Action: transverse (scraping)

Contact material (4-part): 3 *(3-part):* 2

Contact material (specific): wood (hard or dry)

Comments: The tool was deliberately unifacially retouched with an antler billet along the used edge prior to use. The ventral edge-aspect led during use with a transverse action on a moderately hard contact material.

Tool 2

Used area(s): 1

Action: longitudinal (slicing-sawing)

Contact material (4-part): 2 *(3-part):* 2

Contact material (specific): hide

Comments: Smoothing and rounding in area 2 is probably a result of prehension. The tool was used longitudinally on a soft-medium contact material.

Tool 3

Used area(s): 1

Action: transverse (scraping)

Contact material (4-part): 2 *(3-part):* 2

Contact material (specific): hide

Comments: Area 2 has retouch, but this is probably accidental or spontaneous and not related to use. Rounding is visible on the used edge. The ventral edge-aspect led with a transverse action on a soft-medium contact material.

Tool 4

Used area(s): 1-4

Action: boring (perforating-boring)

Contact material (4-part): 2 *(3-part):* 2

Contact material (specific): hide

Comments: Flake scars on area 5 may be the result of prehension during use. The tool was used with a boring action (one-way perforating, twisting motion). The wide (proximal) end of the tool faced the user and the tool was rotated clockwise as the distal end was pushed into the soft-medium contact material.

Tool 5

Used area(s): 1,2

Action: longitudinal (sawing)

Contact material (4-part): 4 *(3-part):* 3

Contact material (specific): antler

Comments: The tool was used longitudinally on a hard contact material.

Tool 6

Used area(s): 1,2

Action: transverse (scraping)

Contact material (4-part): 4 *(3-part):* 3

Contact material (specific): bone or antler

Comments: The tool was either retouched with a hard percussor and not used, or was used transversely on a hard contact material.

Tool 7

Used area(s): 3

Action: longitudinal (slicing)

Contact material (4-part): 1 *(3-part):* 1

Contact material (specific): flesh

Comments: Flake scars on areas 1 and 2 are probably resulted from prehension during use. The tool was used longitudinally on a soft contact material.

Tool 8

Used area(s): 1

Action: longitudinal (slicing-sawing)

Contact material (4-part): 1 *(3-part):* 1

Contact material (specific): flesh

Comments: Flake scars on area 2 may be the result of spontaneous retouch during manufacture. The used edge (area 1) feels dull to the touch and facetting or rounding is suggested under the hand lens. The tool was used longitudinally on a soft contact material.

Tool 9

Used area(s): 1

Action: transverse (scraping)

Contact material (4-part): 4 *(3-part):* 3

Contact material (specific): bone or antler

Comments: The tool was used transversely on a hard contact material with the ventral edge-aspect leading.

Tool 10

Used area(s): 1

Action: transverse (whittling)

Contact material (4-part): 2 *(3-part):* 2

Contact material (specific): wood (soft, fresh)

Comments: The tool was used transversely on a soft-medium contact material.

Blind Test 2

Interpretations for the this test were made on the basis of microflake size, termination type, location and distribution, as well as rounding and polish visible at 40 x with a stereomicroscope. Magnifications up to 100 x were occasionally employed.

Dates Undertaken: July 5,6,12, 1986

Time: total=449 min.; mean=44.9 min.; shortest time=26 min., longest time =69 min.

Tool 1

Used area(s): 1

Action: transverse (scraping)

Contact material (4-part): 3/4 *(3-part):* 2/3

Comments: The tool was percussion retouched in area 1 prior to use. Part of the retouched edge is rounded with an orientation to the ventral (leading) edge-aspect. The pattern of microflake scars is most similar to experimental tools used to scrape hard, dry wood and also somewhat similar to tools used to scrape soaked antler. The tool was used transversely on a medium-hard to hard material. Area 2 also has deliberate retouch, but is probably unused, or only slightly used.

Tool 2

Used area(s): 1

Action: longitudinal (slicing)

Contact material (4-part): 1/2 *(3-part):* 1/2

Comments: Microflake scar patterns are very similar to those on experimental tools used to saw fresh wood, slice dry plants and to cut flesh, although this tool has many more scars. In addition, there is light rounding of a few spots along the edge. The tool was used longitudinally on a soft to soft-medium material.

Tool 3

Used area(s): 1

Action: transverse (scraping)

Contact material (4-part): 2/3 *(3-part):* 2

Comments: Scar patterns are most similar to experimental tools used to scrape bark and wood and also similar to tools used to scrape fresh wood. Slight edge rounding is present, indicating the tool was probably not heavily used. The tool was used with a transverse action (ventral edge-aspect leading) on a medium contact material.

Tool 4

Used area(s): 1-4

Action: boring (perforating/boring)

Contact material (4-part): 2 *(3-part):* 2

Comments: Microflake scar patterns are most similar to those on an experimental tool used to bore soaked antler, although the scars on tool 4 are smaller, suggesting it was used to bore a softer material. The tool functioned as a borer/perforator and was used on a medium or soft-medium contact material.

Tool 5

Used area(s): 1,2

Action: Longitudinal (sawing)

Contact material (4-part): 4 *(3-part):* 3

Comments: Microflake scars are similar in pattern to those on experimental tools used to saw soaked antler or dry wood. The tool was used longitudinally on a medium-hard to hard contact material.

Tool 6

Used area(s): 1

Action: longitudinal (sawing?)

Contact material (4-part): 2 *(3-part):* 2

Comments: Areas 2-4 were deliberately retouched, but unused. Area 1 was used with a longitudinal action on a medium-soft contact material.

Tool 7

Used area(s): 3

Action: longitudinal (sawing)

Contact material (4-part): 2 *(3-part):* 2

Comments: The microflake scar pattern is very similar to that on an experimental tool used to saw birch bark. The tool is interpreted as having been used longitudinally on a medium-soft contact material.

Tool 8

Used area(s): 1

Action: longitudinal (slicing)

Contact material (4-part): 1 *(3-part):* 1

Comments: The microflake scar pattern is most similar to that of experimental tools used to cut dry or green plants, flesh, or very fresh hide and flesh (skinning). The tool was used longitudinally on a soft contact material.

Tool 9

Used area(s): 1

Action: transverse (scraping)

Contact material (4-part): 3 *(3-part):* 2

Comments: Rounding of scar ridges and backs and other high points is extensive on the dorsal edge-aspect. One projecting part of the edge is very rounded. Bright, shiny polish is present on the rounded areas. Both the rounding and the polish are scattered in distribution. The microflake scar pattern is most similar to that of experimental tools used for scraping dry, hard wood, except the scars are considerably larger on this tool. The tool was used transversely (ventral edge leading) on a medium to medium-hard contact material.

Tool 10

Used area(s): 1

Action: transverse (whittling)

Contact material (4-part): 2/3 *(3-part):* 2

Comments: Slight edge rounding is present on one part of the edge and appears to be oriented more to the ventral than dorsal edge-aspect. The microflake scar pattern is most similar to that on experimental tools used to whittle dry wood. The tool was used transversely on a medium to mediumj-hard contact material.

Blind Test 3

Interpretations were made on the basis of micropolish, rounding and striations visible at up to 600 x with a metallographic microscope employing incident lighting. Magnifications used were mainly in the 200 x to 300 x range.

Dates Undertaken: July 15-22, 1986

Time: total=610 min.; mean=61.0 min.; shortest time=20 min., longest time=101 min.

Tool 1

Used area(s): 1

Action: transverse (scraping)

Contact material: wood

Comments: The used edge is rounded and has large patches of very bright, smooth-pitted polish surrounded by bright, grainy polish. This rounded edge is quite broad in places, and patches of bright, smooth polish are sometimes very large. Polish is present only near the edge on the ventral edge-aspect, but occasionally extends a considerable distance back from it, and some large patches of very bright, smooth-pitted polish are present. Bright, but grainy-textured polish is also present on the ventral edge-aspect. Only scattered traces of polish are present on the dorsal edge-aspect immediately above the edge. This polish is mainly bright and grainy and is rarely smooth-textured. Striations are very common in the smooth-pitted polish and invariably run at, or very near, right angles to the edge. These striations are short to medium in length, shallow and narrow.

The tool is interpreted as having been used with a transverse action with the ventral edge-aspect leading. The contact material was wood, as indicated by the polish.

Tool 2

Used area(s): 1

Action: longitudinal (sawing)

Contact material: plant stems

Comments: The used edge is slightly to moderately rounded or blunted. There are spots of very bright, smooth-pitted polish located mainly on a few scattered high points, but especially concentrated on the outward projecting part of the used edge. This polish is surrounded by less bright, grainier polish. On the ventral and dorsal edge-aspects, polish is very extensive and is consistently found on the higher parts of the microtopography, but is not restricted to these areas. Patches of very bright, smooth-pitted polish are often very large, domed, and extend a considerable distance from the edge. Trend lines in the polish are consistently oriented parallel to the edge. In the vicinity of the projecting part of the edge this polish is nearly continuous; elsewhere it is discontinuous to scattered. Striations are not common, but the few which are present are oriented parallel to the edge. These are all narrow, shallow and with the exception of one, very short.

Polish brightness and extent, land the lack of internal grainyness indicates plants are the contact material, and all directional indicators point to a longitudinal action. The tool was used to cut plant stems.

Tool 3

Used area(s): 1

Action: transverse (scraping)

Contact material: wood and bark

Comments: The used edge is slightly rounded to very rounded in places. Rounded parts of the edge have patches of very bright, smooth-pitted polish surrounded by dull to bright, grainy-textured polish. Scattered polishes of the same types found on the edge are found on the ventral edge-aspect near the edge. Polish is much more extensive on the dorsal edge-aspect and extends along ridges and the backs of scars above the edge. Most polish is concentrated on a slightly outward projecting part of the edge. Numerous striations are present on the bright, smooth-pitted polish. These are oriented at or near right angles to the edge and are short to medium in length, narrow and shallow.

The tool was used transversely with the ventral edge leading. The contact material was wood (and bark) as indicated by the polish.

Tool 4

Used area(s): 1-4

Action: boring (perforating/boring)

Contact material: hide

Comments: There are few clear traces of polish on the tool, and edge rounding and striations are absent. There are slight traces of dull, grainy polish on the center ridge of the dorsal face near the tip, and also some suggestions of slightly brighter and smoother patches of polish on various surfaces around the tip.

The tool was not used long enough to develop enough polish to conclusively identify the contact material, however, the traces present suggest hide was the contact material. Tool action is inferred as boring (perforating/boring).

Tool 5

Used area(s): 1,2

Action: longitudinal (sawing)

Contact material: antler/bone

Comments: There are very small patches of bright, smooth polish on a few high points on the used edge, as well as on both edge-aspects. There are also traces of dull, grainy polish. The traces of polish indicate the tool was used with a longitudinal action on either bone or antler, probably to saw through an antler beam.

Tool 6

Used area(s): 4,5

Action: 4 transverse (scraping), 5 graving

Contact material: antler/bone

Comments: Areas 2 and 3 only have traces of dull, grainy polish and if these areas were used, they were not used long enough for a distinctive polish to develop. Area 4 has the slightest traces of dull, grainy polish on the edge and dorsal edge-aspect, as well as a few very small patches of smooth, bright-pitted polish. No polish is present on the ventral edge-aspect. Several possible striations running at right angles to the edge were present in the polish on the edge. Area 5 has three very small patches of bright, smooth polish surrounded by dull, grainy polish. Striations are also present and these run parallel to the long axis of the spur.

None of the areas defined for the tool were used long enough for clear-cut wear patterns to develop and if this was a prehistoric tool I would interpret it as used, period. My forced interpretation is: area 1 was not used; areas 2 and 3 were probably used, but the traces are not distinctive enough to determine contact material; area 4 was used transversely on a hard contact material, probably antler or bone (hard, dry wood?). Area 5 may have been used with a graving action and the polish is probably from use on antler or bone (wood?).

Tool 7

Used area(s): 3

Action: longitudinal (slicing)

Contact material: flesh and fat

Comments: No definite polish or striations are present on the tool. In area 3 there are possible traces of bright, rough-textured polish and some slight smoothing of high points on both edge-aspects extending a considerable distance back from the edge. There are also slight suggestions that area 2 may have also been used, but not enough to say conclusively. The traces there may be the result of prehension.

The edge was not used long enough for a distinctive polish to develop and if this was a prehistoric tool I would be reluctant to assign a function on the basis of such meager wear traces. The smoothing suggests deep penetration into a contact material, proably a soft one. The tool may have been used longitudinally on flesh and fat (or hide?).

Tool 8

Used area(s): 1

Action: longitudinal (slicing-sawing)

Contact material: flesh and fat

Comments: Polish and striations are absent from the edges and edge-aspects. In area 1, the part of the edge with microflake scars, there is the slightest hint of edge rounding. More significant is an apparent slight smoothing of high points on the edge-aspects extending a considerable distance from the edge. This indicates deep penetration into a soft contact material.

Again, I am reluctant to assign a function to this tool based on very slight wear traces. I would have to suggest it was used longitudinally on flesh and fat (or hide?), but probably for only a short time.

Tool 9

Used area(s): 1

Action: transverse (scraping)

Contact material: wood

Comments: There are only a few locations along the edge where rounding is present and polish is well-developed. At three places along the edge very bright, smooth-pitted polish is present. This is surrounded by bright-grainy polish, which, along with dull-grainy polish is also present elsewhere along the edge. Traces of bright, smooth polish are present on two locations on the ventral edge-aspect. On the dorsal aspect, bright-grainy and dull-grainy polishes are relatively common on ridges and high points, along with traces of bright, smooth-pitted polish. Striations are common in the bright, smooth polish and run at right angles to the edge. They are short to medium in length, narrow and shallow.

The tool was used with a transverse action with the ventral edge-aspect leading. The contact material was wood as indicated by the polish. The wood was probably very hard or very dry, which caused the removal of many microflakes, and at the same time most of the polish as it was developing.

Tool 10

Used area(s): 1,2

Action: transverse (whittling)

Contact material: wood

Comments: The edge has only slight rounding and small patches of smooth, bright-pitted polish, but more bright, grainy and dull, grainy polishes. On the ventral edge-aspect, polish is very extensive and consists mainly of very bright, smooth-pitted polish concentrated next to the edge, especially along the projecting part of the edge. The dorsal edge-aspect has the least amount of polish. There are numerous striations on the bright, smooth polish on the ventral edge-aspect. These are short to medium in length, narrow, shallow and oriented at or near right angles to the edge.

The polish indicates wood is the contact material and all directional indicators point to the tool having been used with a transverse action. The minimal degree of edge rounding and its distribution suggest the tool was used with a whittling action.

www.ingramcontent.com/pod-product-compliance
Ingram Content Group UK Ltd.
Pitfield, Milton Keynes, MK11 3LW, UK
UKHW061845041026
15591UKWH00015B/925

9 780860 545910